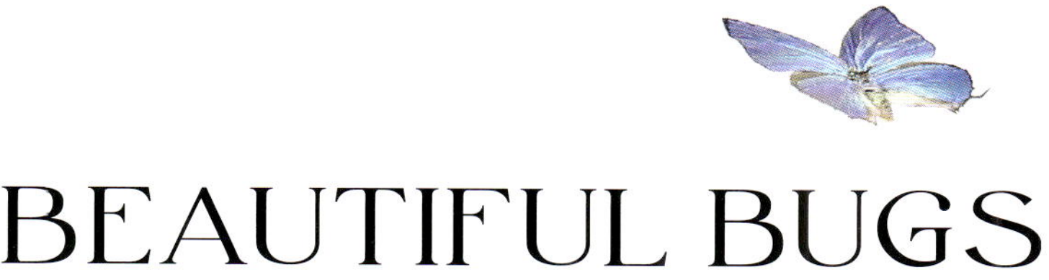

BEAUTIFUL BUGS

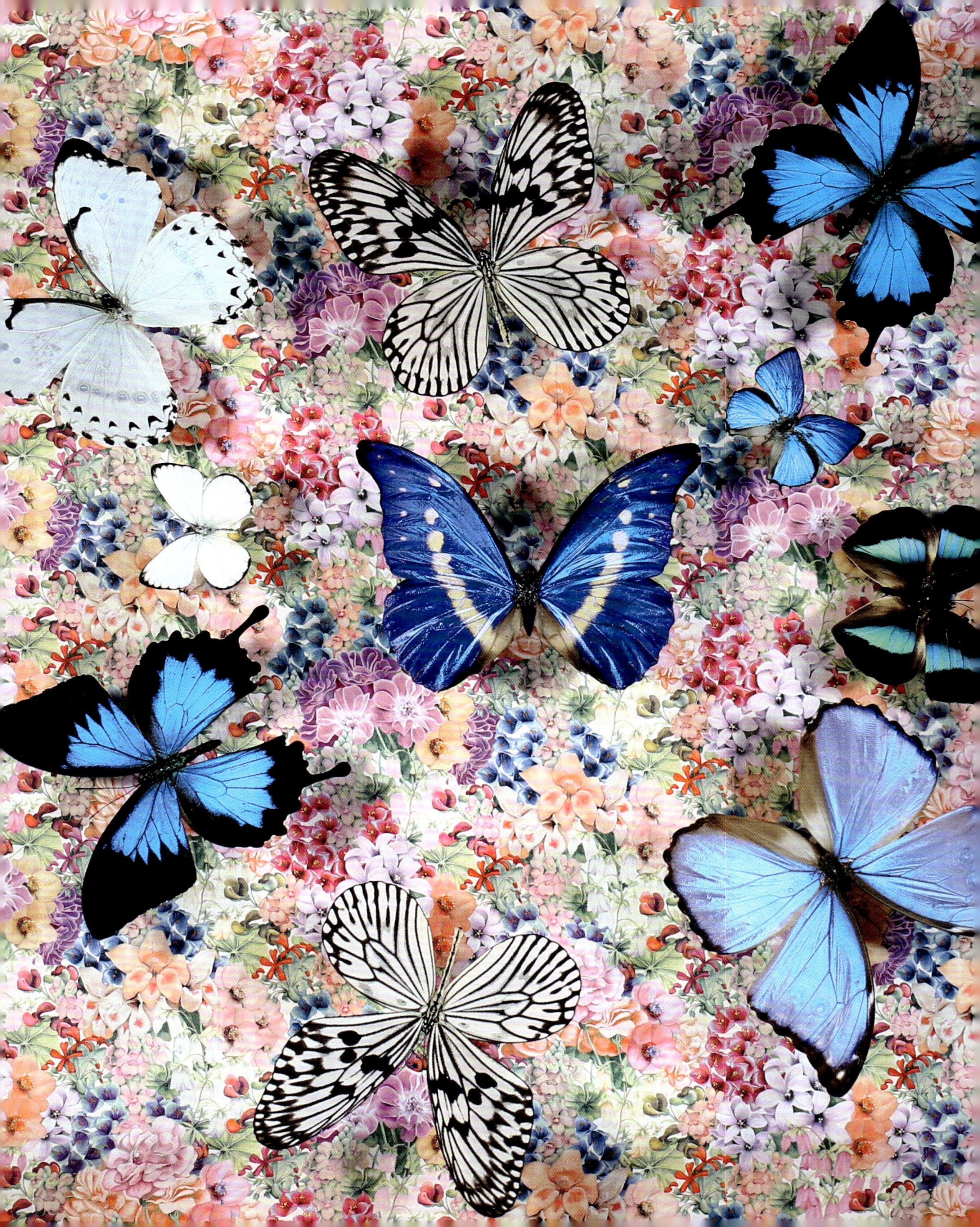

BEAUTIFUL BUGS

A Curious Collection of Exquisite Insects

By Bug in the Box

CHRONICLE BOOKS

SAN FRANCISCO

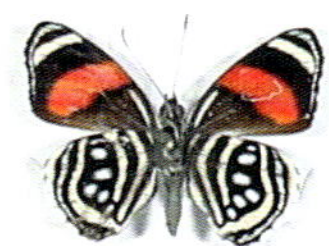

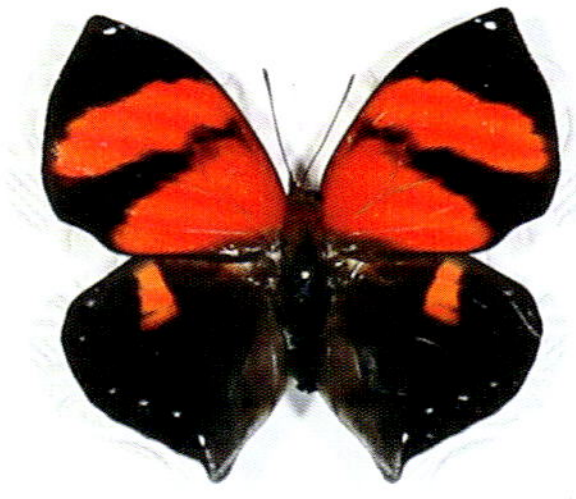

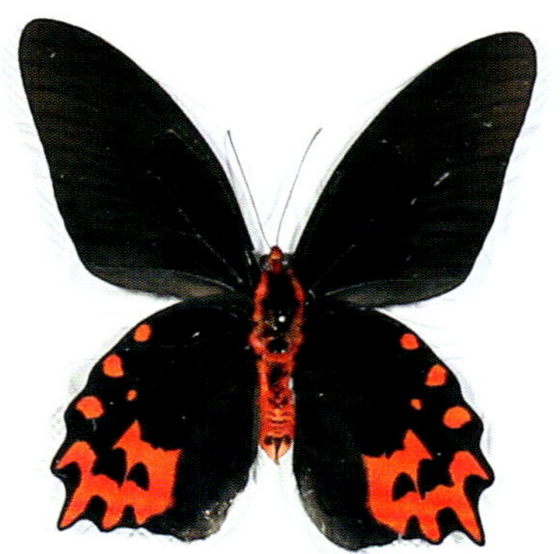

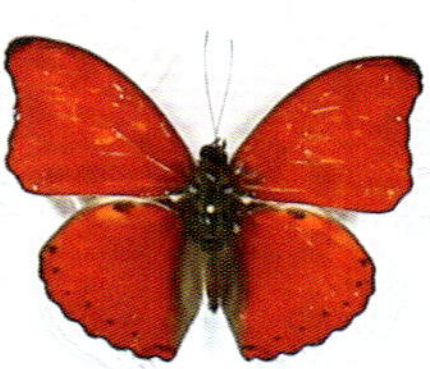

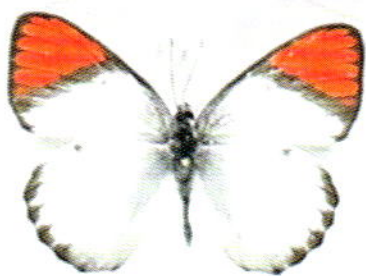

Library of Congress Cataloging-in-Publication Data available.

ISBN 978-1-7972-4304-7

Manufactured in China.

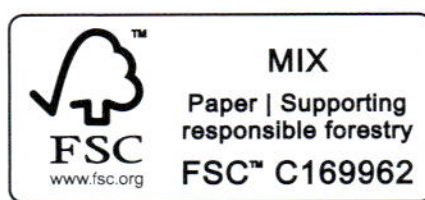

Design by Kayla Ferriera.

10 9 8 7 6 5 4 3 2 1

Chronicle books and gifts are available at special quantity discounts to corporations, professional associations, literacy programs, and other organizations. For details and discount information, please contact our premiums department at corporategifts@chroniclebooks.com or at 1-800-759-0190.

Chronicle Books is represented in the UK and Europe by Abrams & Chronicle Books, 22 Ely Place, London EC1N 6TE, abramsandchronicle.co.uk, and Média-Participations, 57 rue Gaston Tessier, 75166 Paris CEDEX 19, media-participations.com.

Chronicle Books LLC
680 Second Street
San Francisco, California 94107
chroniclebooks.com

CONTENTS

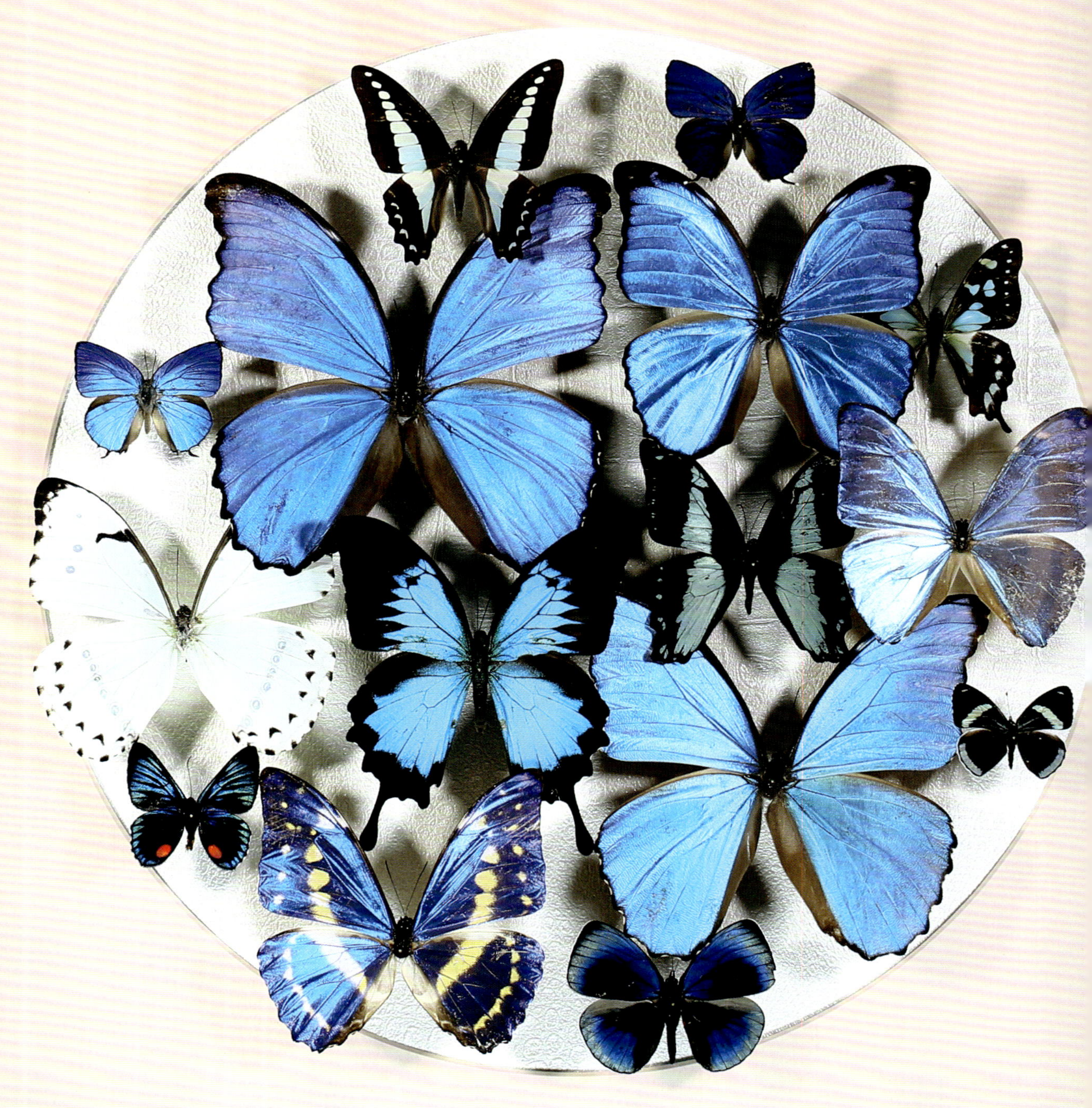

Introduction

Welcome to *Beautiful Bugs*, a collection of some of the world's most fascinating insects. You may be here as a bug enthusiast with an established knowledge of these creatures, or you may be a curious passerby. Regardless, prepare to have your breath stolen as you explore these pages.

This book is a testament to years of dedicated artistry and thorough scientific research, seamlessly woven together to create unique, artfully crafted and preserved insect pieces. My partner, Tristan, and I carefully design each piece, showcasing the insects in exceptional ways. We source vintage fabrics and exquisite papers; preserve and dye ferns and flowers; and collect antique frames to beautifully display these interesting creatures. Every element is chosen to complement and enhance the striking features of the insects. Our work is inspired by a diverse array of influences, from the charm of old curios to the sleek simplicity of modern minimalist design. We create pieces that cater to a variety of tastes, each one distinct in hopes that our work will resonate with anyone who appreciates beauty in nature and art.

The collection in this book spans from our humble beginnings in 2015 to the present day. It traces our evolution as artists and reflects our enduring passion for this craft. What began as personal gifts for friends and family quickly garnered admiration, prompting us to share our creations with a wider audience. Encouraged by their support, we began participating in art shows and pop-up galleries, slowly building a following on social media. Today, we are proud to have a dedicated studio and showroom in Houston, as well as a permanent storefront in the charming town of Round Top, Texas.

Coming from a family steeped in both entomology and art, I'm lucky to have a strong and inspired foundation for this work. My mother, a passionate entomologist with a particular focus on Mexico's diverse ecosystems, introduced me to the delicate art of preserving and pinning insects. As a child,

I often accompanied her on specimen-collecting trips during our vacations to Mexico, where we would explore lush forests and habitats teeming with insects of all kinds. These excursions opened my eyes to the intricate details of the natural world. My mother would share her vast knowledge of the local species, explaining their diets, habitats, and ecosystems. Some of my most cherished memories are of us hanging a large white sheet by the side of our house and shining a bright UV light to attract moths, katydids, and other nighttime insects. Under her guidance, I was introduced to creatures I otherwise never would have encountered, and she would patiently teach me about them, sparking a lifelong fascination.

One of the most valuable gifts she gave me was her expertise in preserving insects. She created a special mixture of oils and alcohol that kept insects vibrant and intact for generations, and it's the same mixture that we use today for our pieces. Although she has since passed away, her teachings continue to shape the work I do. Every piece I create carries a part of her—her passion, her wisdom, and the skills she imparted to me.

Growing up surrounded by art, nature, and insects has instilled in me a deep appreciation for the remarkable diversity of life on our planet. Our firm ensures that all the insects used in our pieces are ethically and humanely sourced. We work closely with farms in Costa Rica and Indonesia, many of which were established through my mother's connections; other partnerships we cultivated ourselves. These farms specialize in butterfly breeding for repopulation efforts. Half of the butterflies raised are released back into the wild, while the other half are kept for breeding the next generation, ensuring a sustainable cycle. We also collaborate with renowned institutions like the Houston Museum of Natural Science, which provides us with insects that have passed away naturally, preserving their legacy in our work.

My hope is that as you journey through the pages of this book, you develop a greater appreciation for the often-overlooked insects that share our planet. These creatures—so diverse and complex—are marvels in their own right, each one playing a crucial role in the natural world.

Insects are true masterpieces of evolution, from the majestic butterflies, with their breathtaking colors and intricate patterns—such as the morphos, whose vivid blue wings reflect specific light wavelengths, creating a mesmerizing iridescent effect—to the elusive beetles that work quietly beneath fallen logs and tree bark, many of which boast the most stunning hues and textures found in nature.

Insects are radically different from humans, yet their intriguing behaviors, delicate life cycles, and delightful quirks are exactly what have drawn me to them. Their lives, often hidden from view, are filled with mystery and wonder, and it is this blend of the extraordinary and the everyday that makes them so captivating. Through these pages, I hope to share not just the beauty but also the profound importance of these small yet mighty creatures.

—Ruben Salazar
CO-OWNER, BUG IN THE BOX

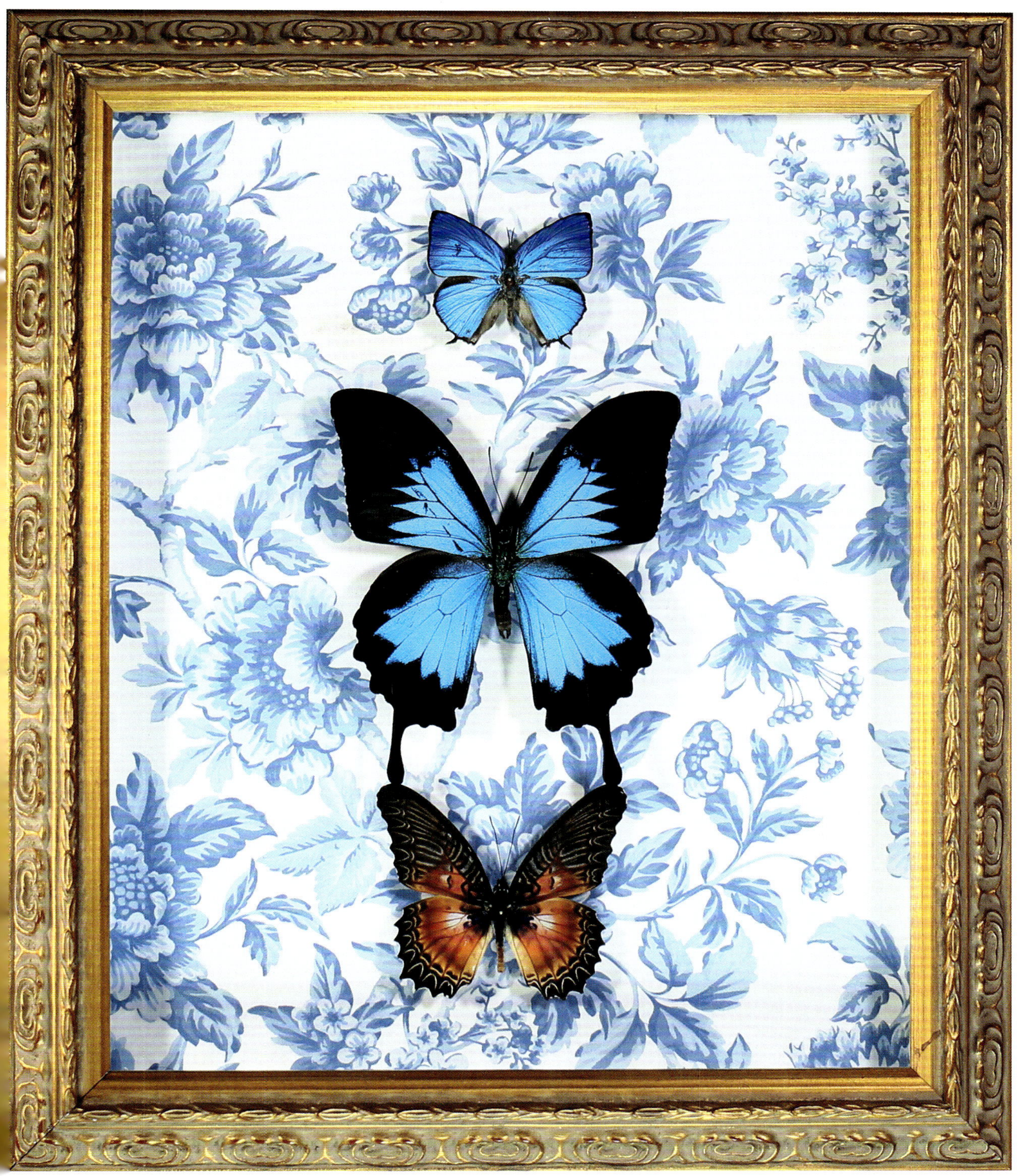

BUG MEANINGS

The creatures presented here offer a glimpse into how insects have been perceived by different cultures around the globe. This sampling includes stories and interpretations that are most resonant to me— they represent a segment of the rich and varied tapestry of human understanding throughout history. Each type of insect signifies basic characteristics of life in the eyes of their human beholders.

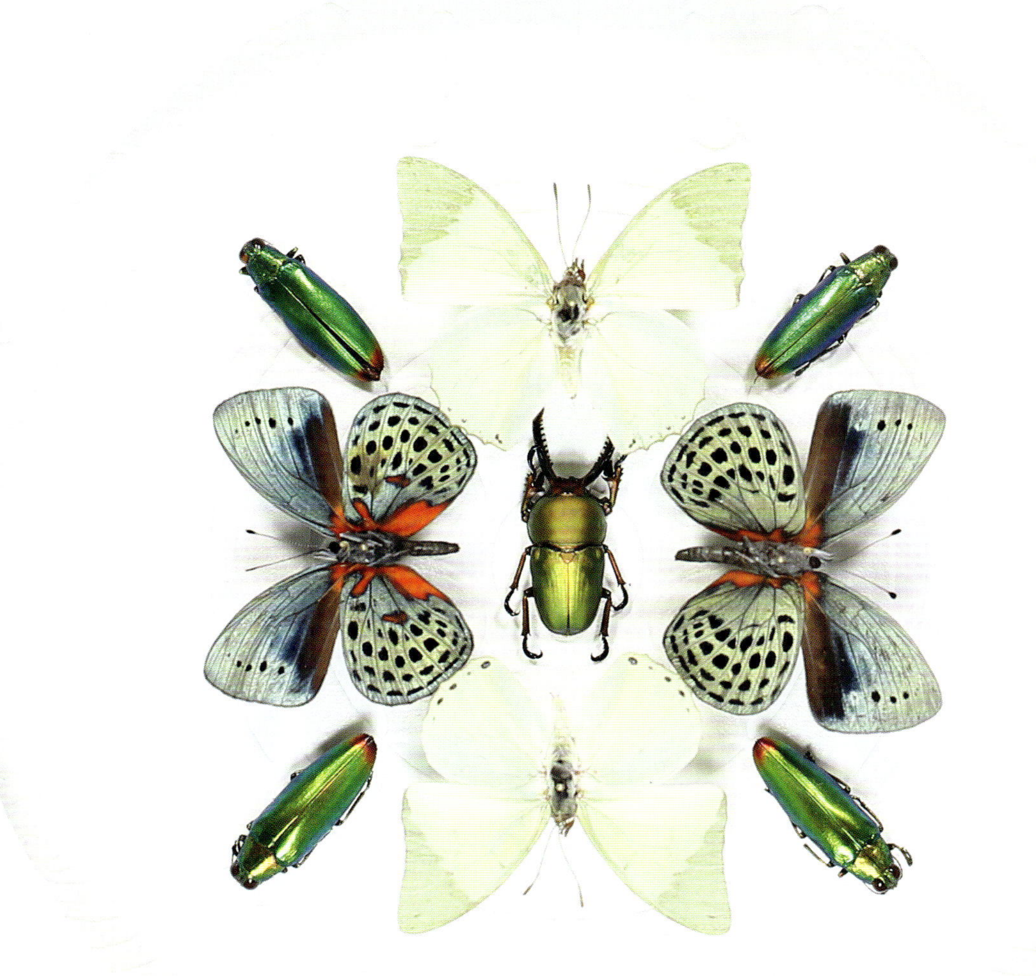

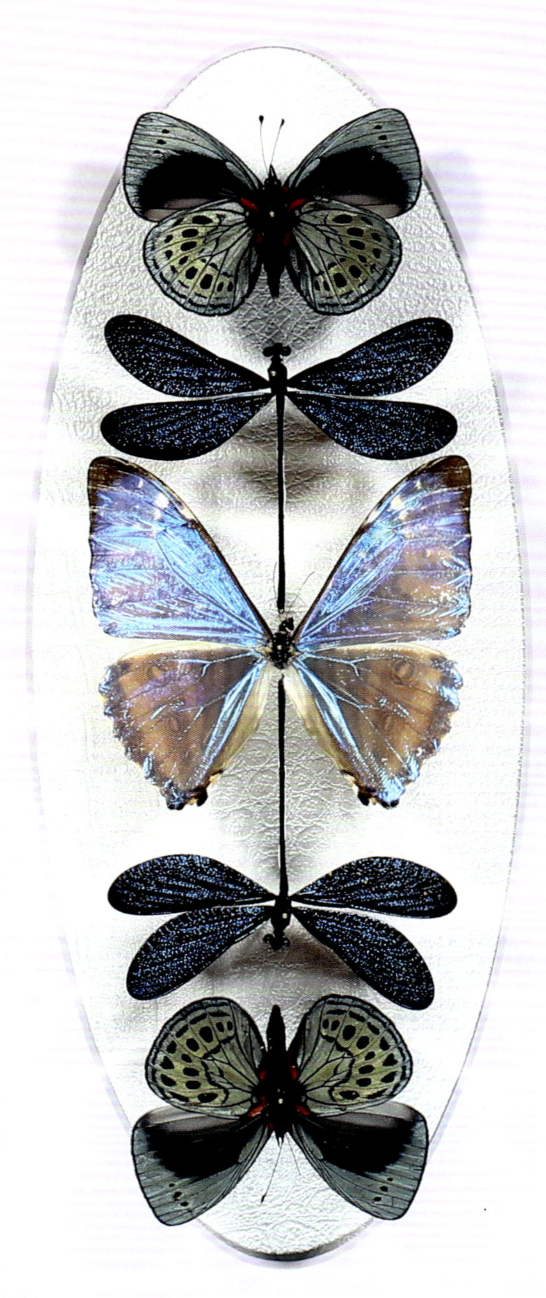

BUTTERFLIES Hope, Change, Spirit

The butterfly's symbolism is tied to its metamorphosis, the process by which it undergoes a complete transformation from one form to another. The first sighting of butterflies in spring is a powerful symbol of hope across many cultures, heralding the renewal of life after the long, harsh winter months. In ancient Greece, the word for butterfly also meant "soul" or "mind," indicating a deep connection between these creatures and the soul's path toward enlightenment and self-discovery. In other cultures, including that of ancient Egypt, butterflies were seen as representations of the enduring spirits of departed loved ones and served as a heartfelt reminder of the connections and memories they left behind. In this way, the butterfly is not only a symbol of change but also a living bridge between the past and the present, carrying with it the essence of those who came before us.

MOTHS Transformation, Death, Determination

Moths symbolize transformation and resilience as they navigate through distinct life stages: transitioning from caterpillars to pupae, emerging as graceful winged beings, and facing death. Their journey serves as a reminder of the importance of stepping beyond our comfort zones to grow, evolve, and realize our fullest potential. In sub-Saharan Africa, moths have been interpreted as the souls of deceased people delivering messages or warnings. In Japan, moths are seen as emblematic of the double-edged sword of beauty and vulnerability. These creatures teach us about the power of change and the wisdom of trusting our instincts as we traverse our own transformative paths.

DRAGONFLIES Prosperity, Courage, Agility

In Japan, as well as in some Native American cultures, the sight of dragonflies hovering over a body of water is considered good luck, suggesting that the area is rich in resources and low in pests. In ancient Japan, dragonflies were revered as symbols of courage and agility, appearing in literature and art and decorating the helmets of Samurai warriors. More broadly, witnessing a swarm of these captivating insects was often seen as an auspicious sign, signaling the arrival of rain and a bountiful harvest. Dragonflies embody abundance and represent the delicate balance and fertility that sustain life.

BEETLES Rebirth, Good Fortune

Beetles have appeared in countless depictions throughout human history and are featured in wide-ranging beliefs about the world. In ancient Egypt, the scarab beetle symbolized the eternal cycle of life, death, and rebirth. Revered for its connection to the sun god Khepri, who was often depicted as a scarab rolling the sun across the sky, the beetle embodied the daily resurrection of the sun and its relentless journey through the heavens. The beetle's role in this cosmic order made it an emblem of both spiritual renewal and the enduring power of life's continual flow. In Chinese tradition, these creatures symbolized wealth and good luck, as their hardened exteriors were sure to protect them from misfortune.

BEES & WASPS Order, Organization, Productivity

From Sumer and Egypt to Greece, India, and medieval Europe, the symbolism of wasps and bees is rooted in their extraordinary social structures and highly organized hives—something that few other creatures parallel. In these colonies, every individual, from the soldiers and workers to the queen, plays a crucial and distinct role, working in concert to maintain the harmony of the group. This intricate division of labor ensures the colony's survival, productivity, and flourishing. As such, bees and wasps are often seen as symbols of cooperation, industry, and collective strength, embodying the power of unity and the importance of each individual's contribution to the greater whole.

PRAYING MANTISES Meditation, Patience, Clarity

Due to the impressive hunting strategy of mantises—where they remain perfectly still for long stretches, patiently waiting to ambush their prey—many cultures across the world view these insects as symbols of meditation and patience. Their stillness and focus serve as powerful metaphors, reminding us of the value in pausing to reflect, sharpen our concentration, and embark on journeys of self-discovery. In some cultures, like that of ancient Greece, mantises were also seen as oracles or travel guides. Their presence imbued observers with clarity about where to go next.

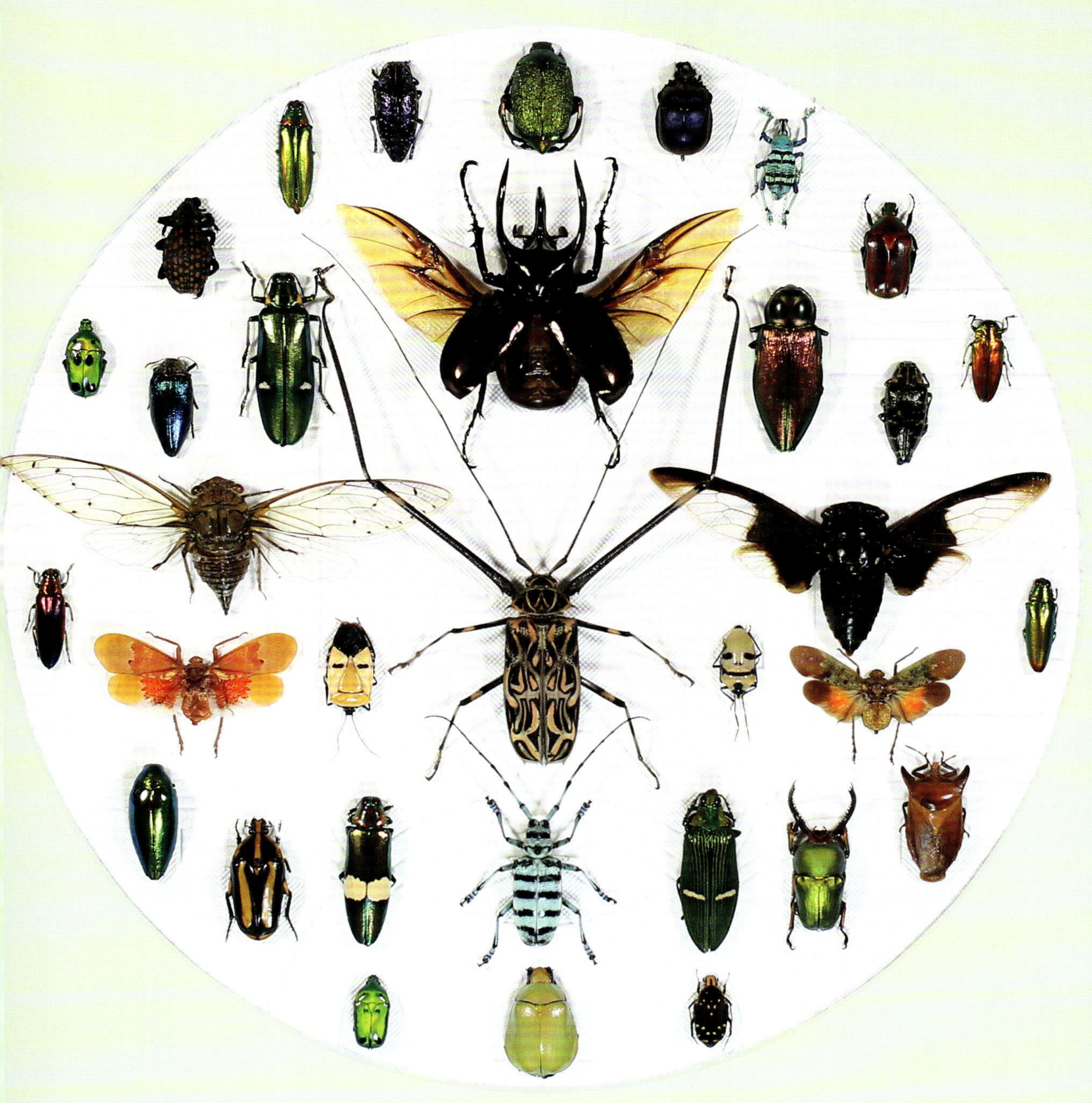

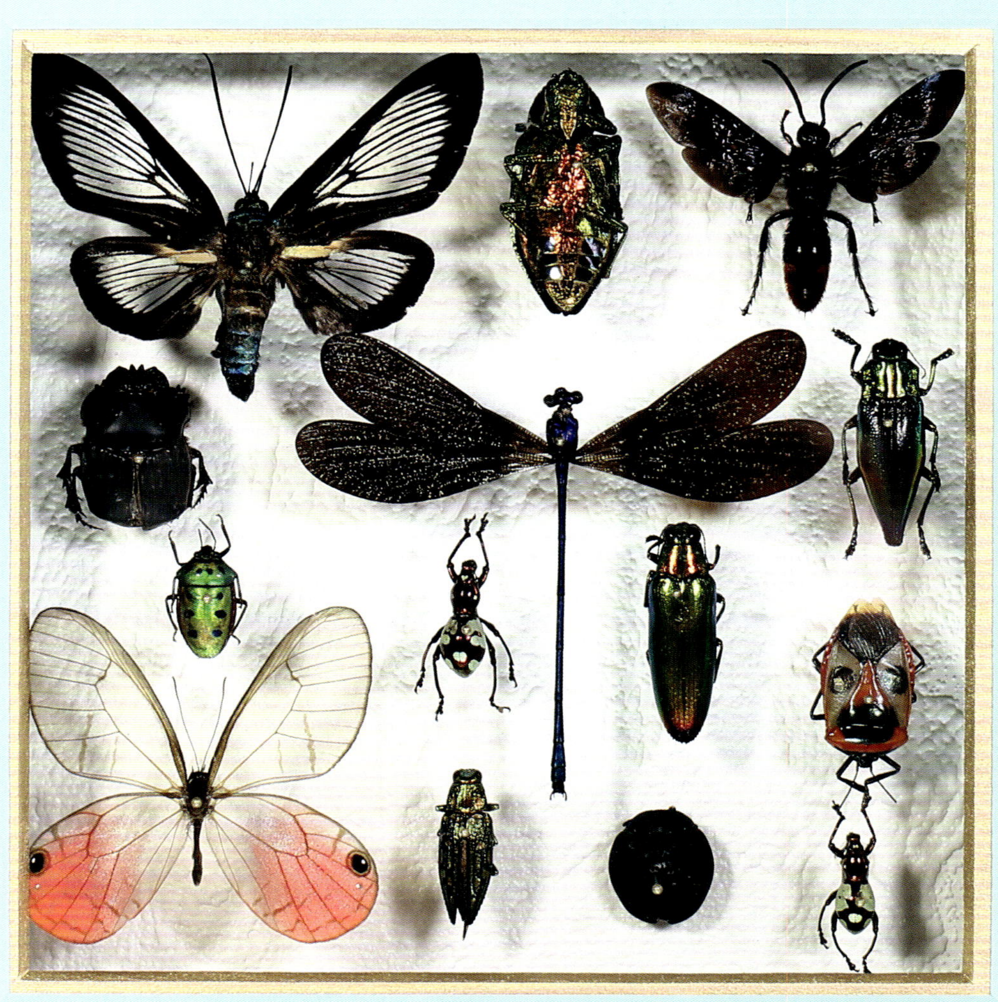

SCORPIONS Passion, Protection, Transition

In many cultures, from that of ancient Egypt to that of modern-day Turkey, scorpions are powerful symbols of passion and protection, known for their swift, fierce defense when threatened. Their ability to strike quickly with a venomous sting has long been associated with intense emotions and the need to guard what is most dear. Scorpions shed their exoskeletons as they grow, making them a symbol of the ongoing cycles of personal evolution. When a scorpion appears in your life, it can be interpreted as a sign that a major transformation is at hand—one that calls for reflection, introspection, and the courage to embrace change.

SPIDERS Creativity, Strength, Balance

In cultural myths and oral traditions from West Africa to the Southwestern United States, a spider's ability to weave is linked to creating fates, holding wisdom, and telling stories. The fine yet incredibly resilient structure of a web reflects the spider's craftsmanship and its ability to capture prey many times its own size. The spider's web serves as a reminder that, like the web builder, we have the ability to weave our own realities through mindful choices and purposeful actions. Each strand, carefully placed, can lead to the creation of something much larger.

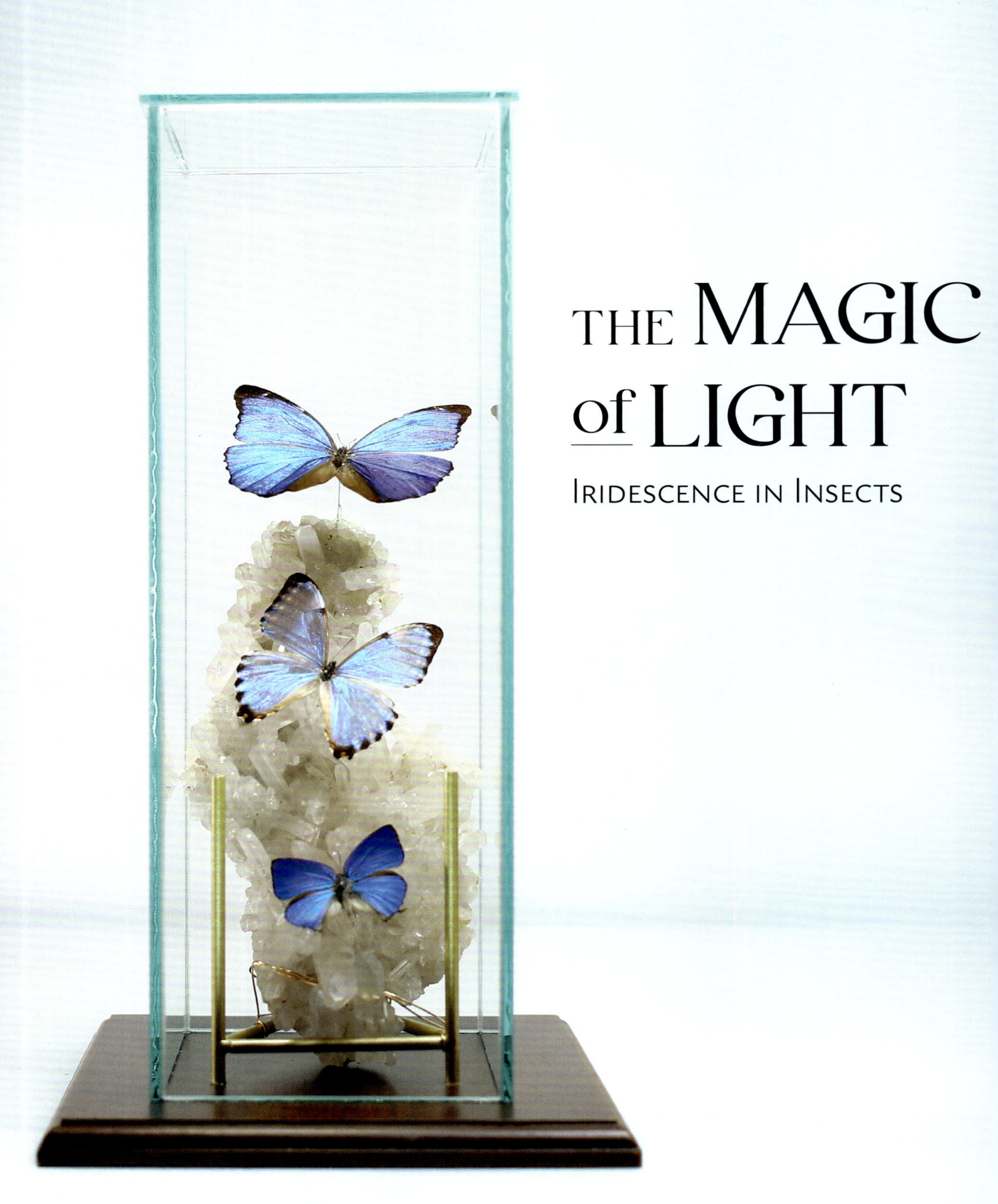

THE MAGIC of LIGHT

IRIDESCENCE IN INSECTS

IRIDESCENCE IS A CAPTIVATING PHENOMENON THAT APPEARS TO BEND the rules of light itself, transforming tiny insects into vibrant, living rainbows. When sunlight strikes the delicate wings of a dragonfly or glistens on the shell of a beetle, colors shimmer and shift in an enchanting manner, creating a dazzling display of hues that wouldn't otherwise appear. The surfaces of iridescent insects are coated with intricate microscopic layers, or specialized nanostructures, that effectively reflect and refract light, allowing us to perceive them as ever-changing and dynamic. As the angle of our view alters, so, too, does the color, producing a mesmerizing effect that seems to pulse with life. It's an optical illusion of sorts, but one rooted in precise biological engineering—a marvel that nature has perfected over millions of years.

Insects harness this shimmering quality in a plethora of interesting ways. For some species, it serves to attract potential mates—consider the iridescent sheen that graces the wings of butterflies or the lustrous armor of certain beetles, both reflecting the individual's health and vitality. For other species, this exquisite visual feature functions as an effective defense mechanism, flashing colors that confuse or startle potential predators. Iridescence can even be strategically employed to achieve camouflage, mimicking the play of light on water or through foliage.

From its role in mating displays and visual signals to sophisticated camouflage and crucial survival tactics, iridescence in insects is a striking reminder of the intricate, often hidden beauty of the natural world. It's a profound story told in the language of light, color, and the dance between biology and the laws of physics that govern our universe.

ZEPHYRITIS MORPHO

SCIENTIFIC NAME
Morpho zephyritis

GEOGRAPHIC RANGE
Brazil, Peru, and Bolivia

The name Zephyritis is derived from Zephyrus, the Greek god of the west wind. Therefore, the common name Zephyritis morpho can be translated as something like "the gentle breeze morpho," speaking to the butterfly's graceful movement through the air.

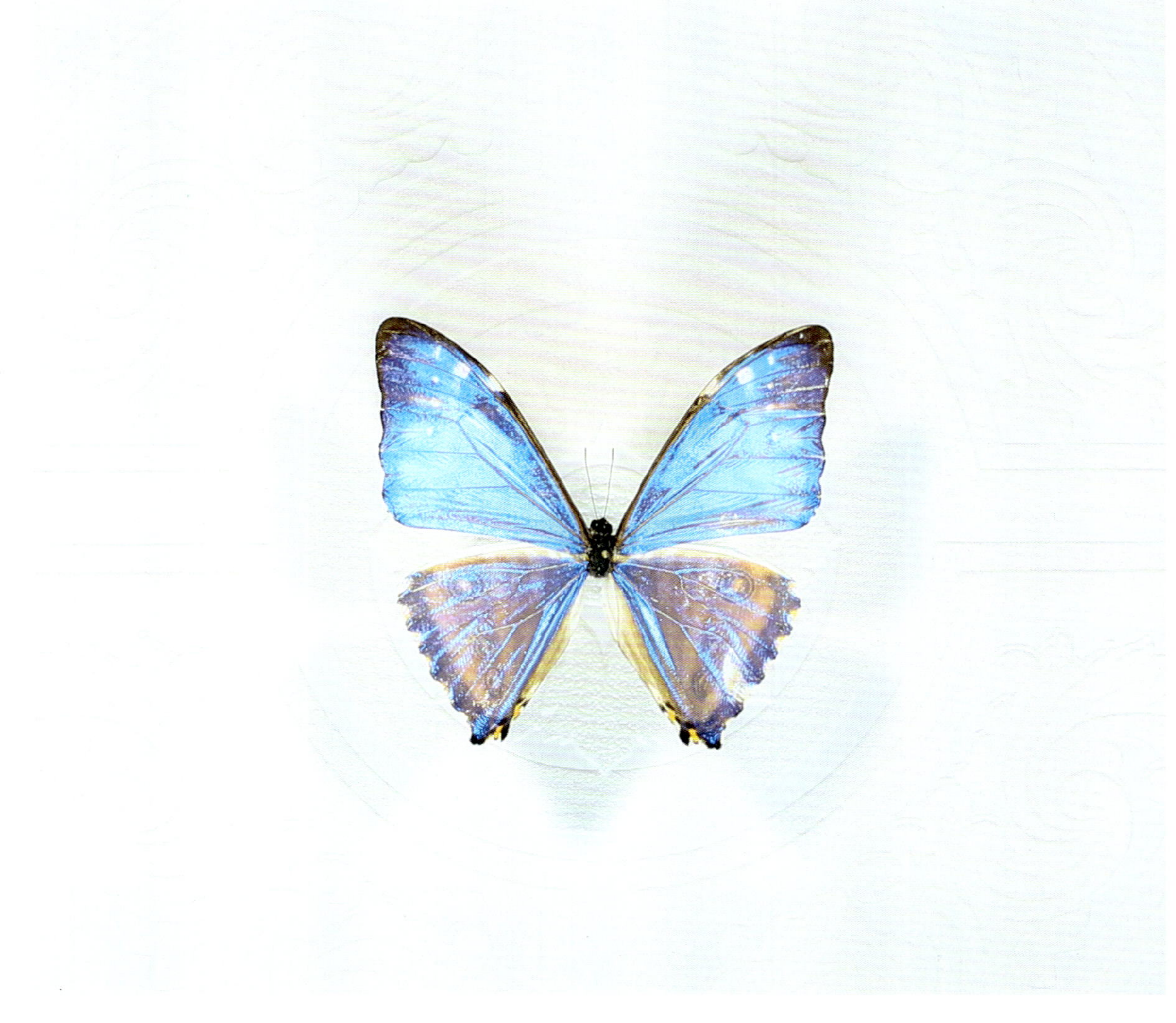

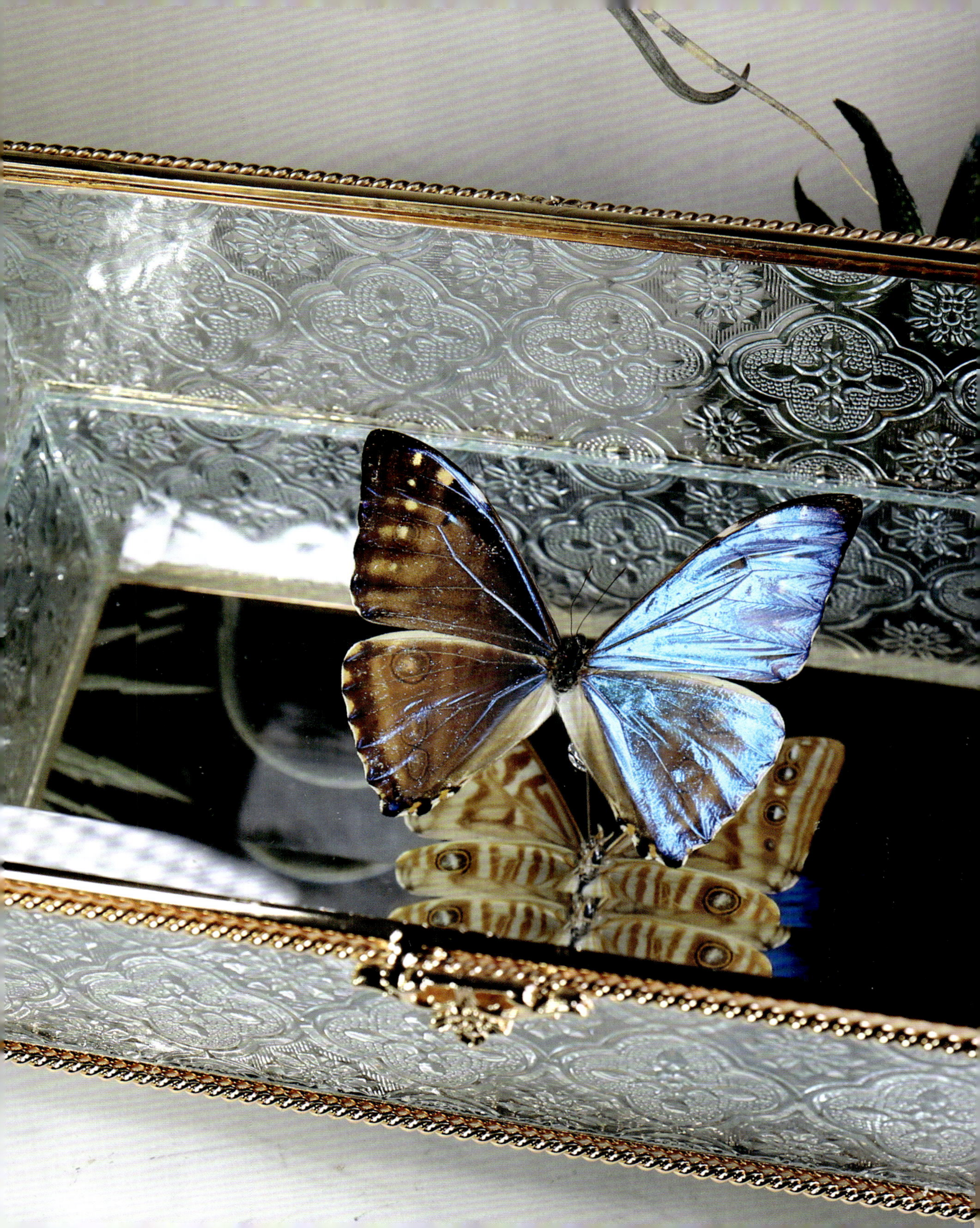

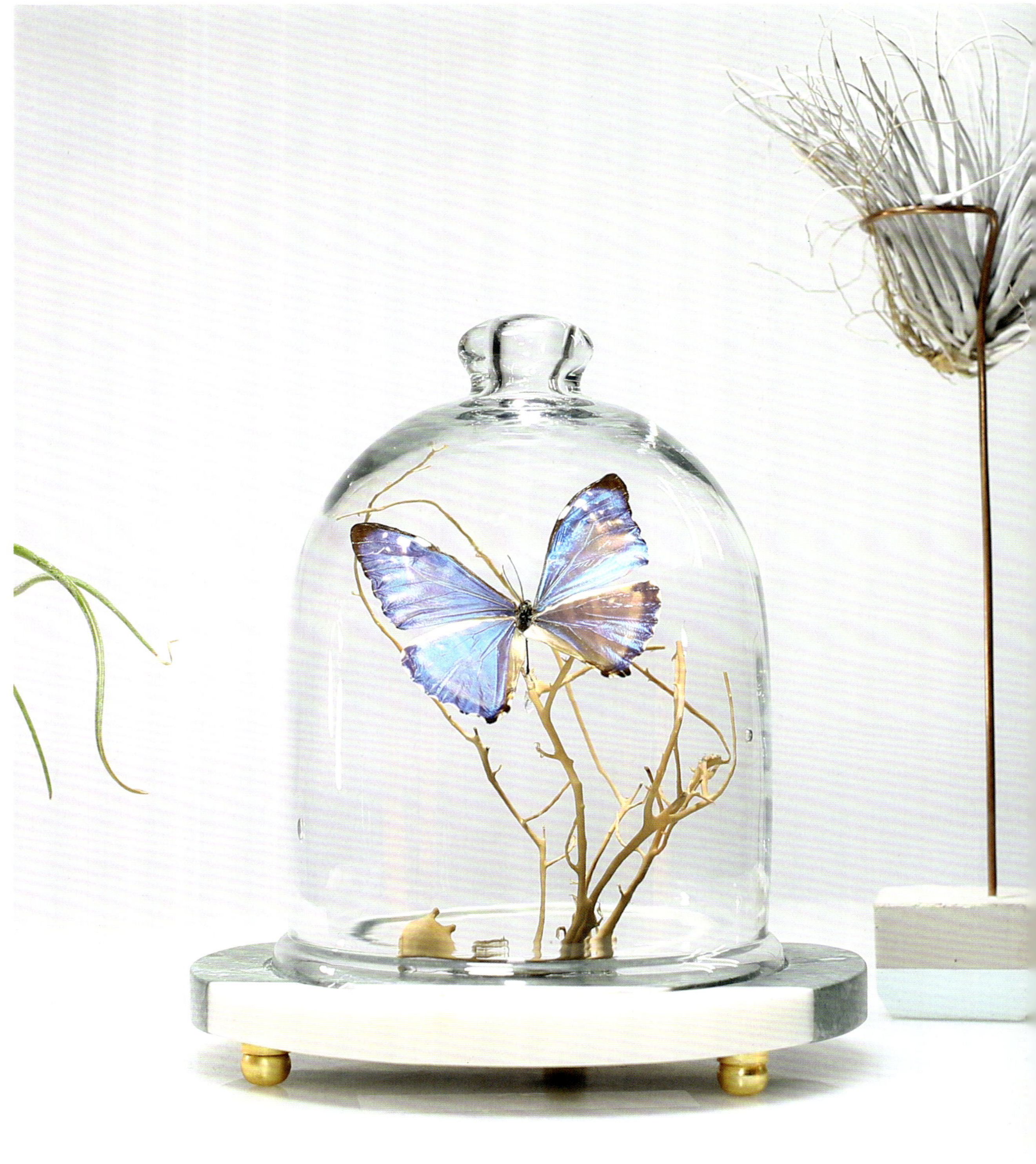

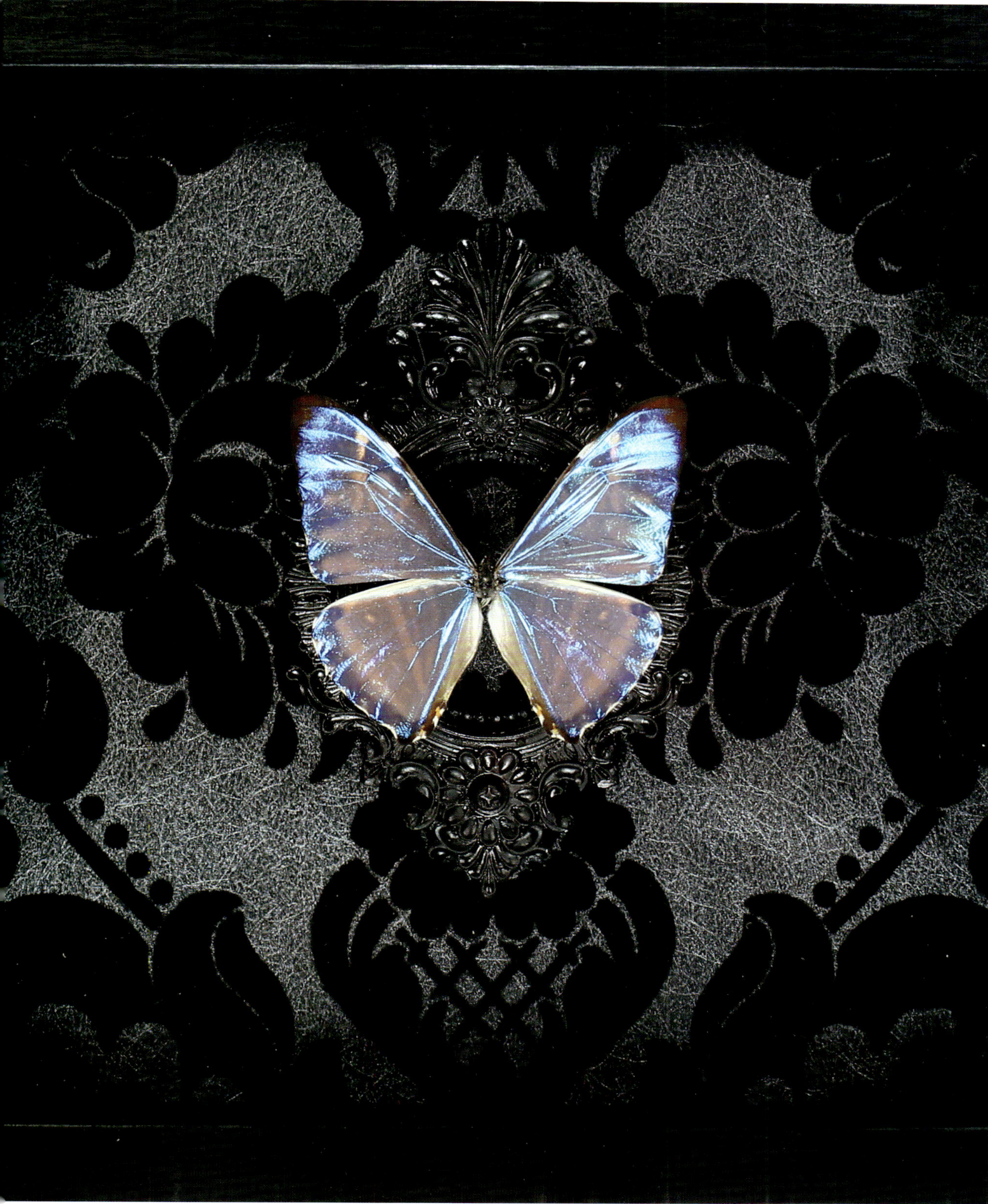

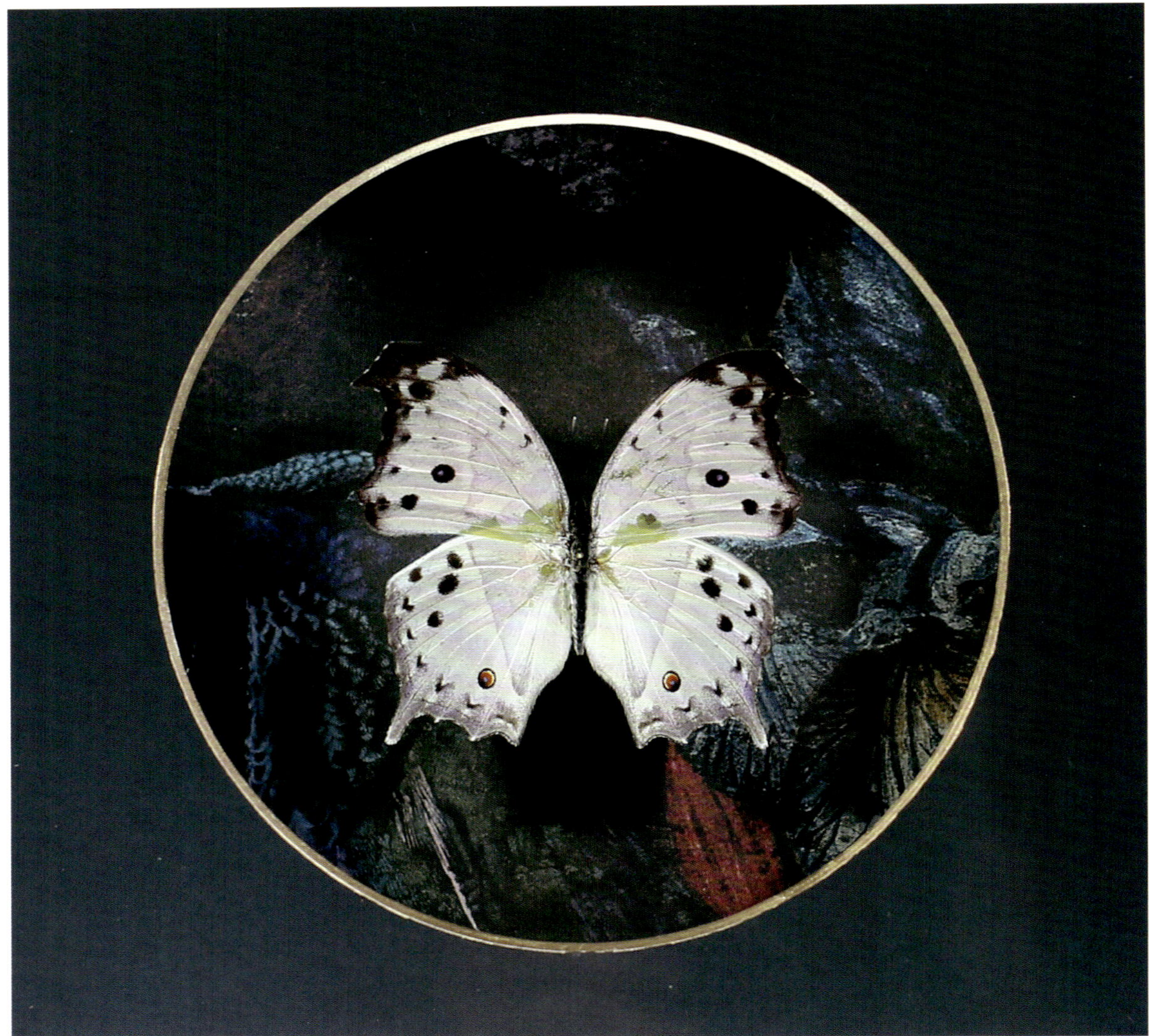

FOREST MOTHER-OF-PEARL

SCIENTIFIC NAME
Protogoniomorpha parhassus

GEOGRAPHIC RANGE
Africa

Female mother-of-pearl butterflies typically dwell in the understory of forests, near plants with larval food supply, while males have been observed atop the trees. This butterfly has a rainbowlike iridescent shine on its delicate wing scales. Its shimmering colors range from soft pastel green to vibrant, eye-catching pink, creating a mesmerizing display that is sure to delight those who have the privilege of observing it.

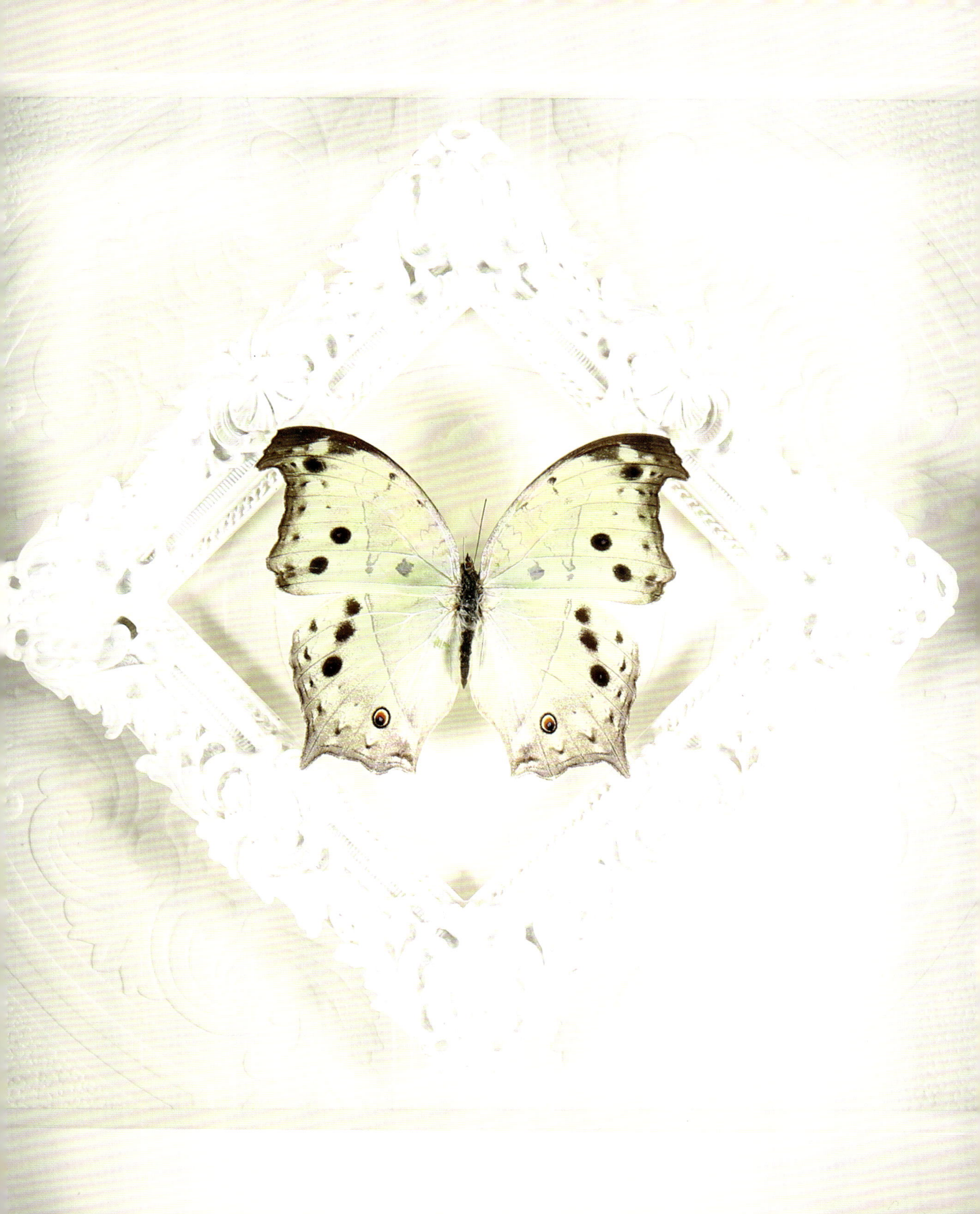

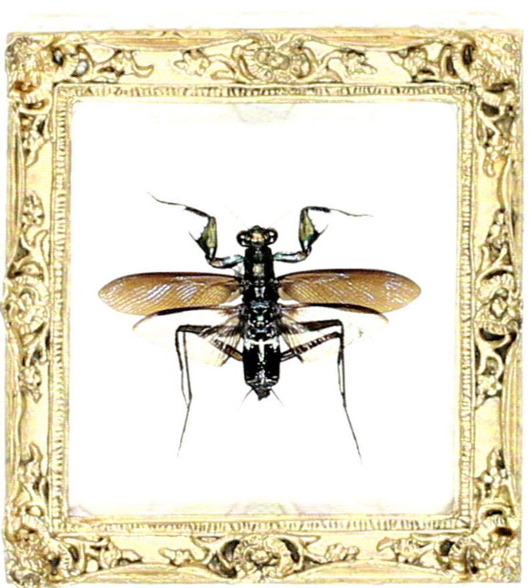

IRIDESCENT BARK MANTIS

SCIENTIFIC NAME
Metallyticus splendidus

GEOGRAPHIC RANGE
Southeast Asia

This mantis is a skilled hunter both during the day and throughout the night. Males and females of this intriguing species exhibit distinct differences in coloration. Males boast a prominent blue-violet hue, while females display a beautiful golden-green shade. Unlike many species of mantises, *M. splendidus* can typically be found hiding under the bark of dead trees.

GIANT BLUE MORPHO

SCIENTIFIC NAME
Morpho didius

GEOGRAPHIC RANGE
Peru

The giant blue morpho's vivid, iridescent blue coloring is a result of the microscopic scales arranged on the backs of its wings, which reflect light beautifully. The underside of the morpho's wings, however, is a decidedly dull brown adorned with eyespots, which provide effective camouflage when its wings are closed.

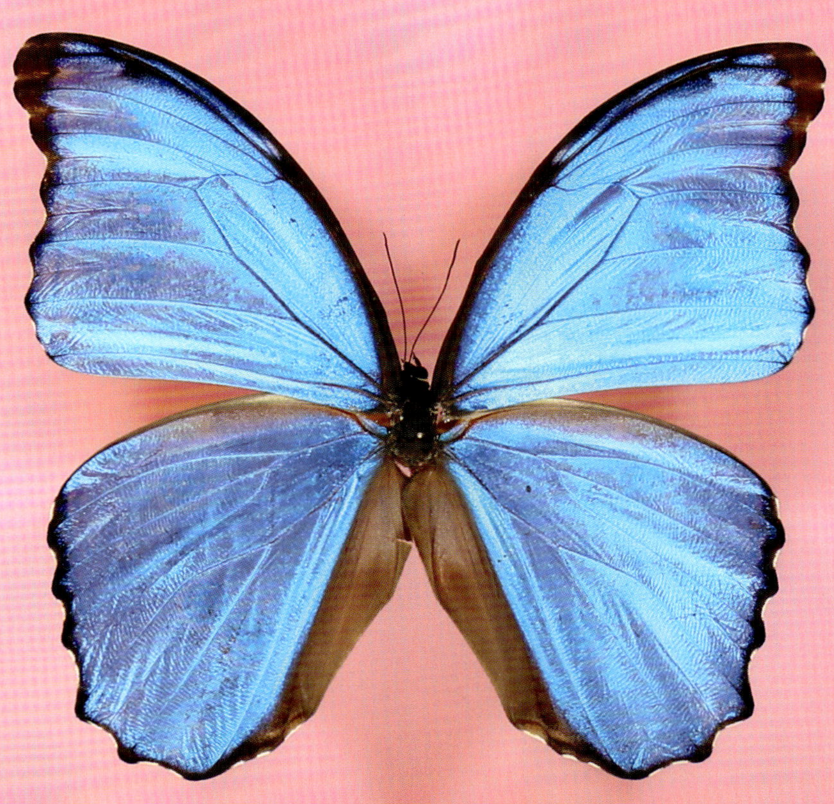

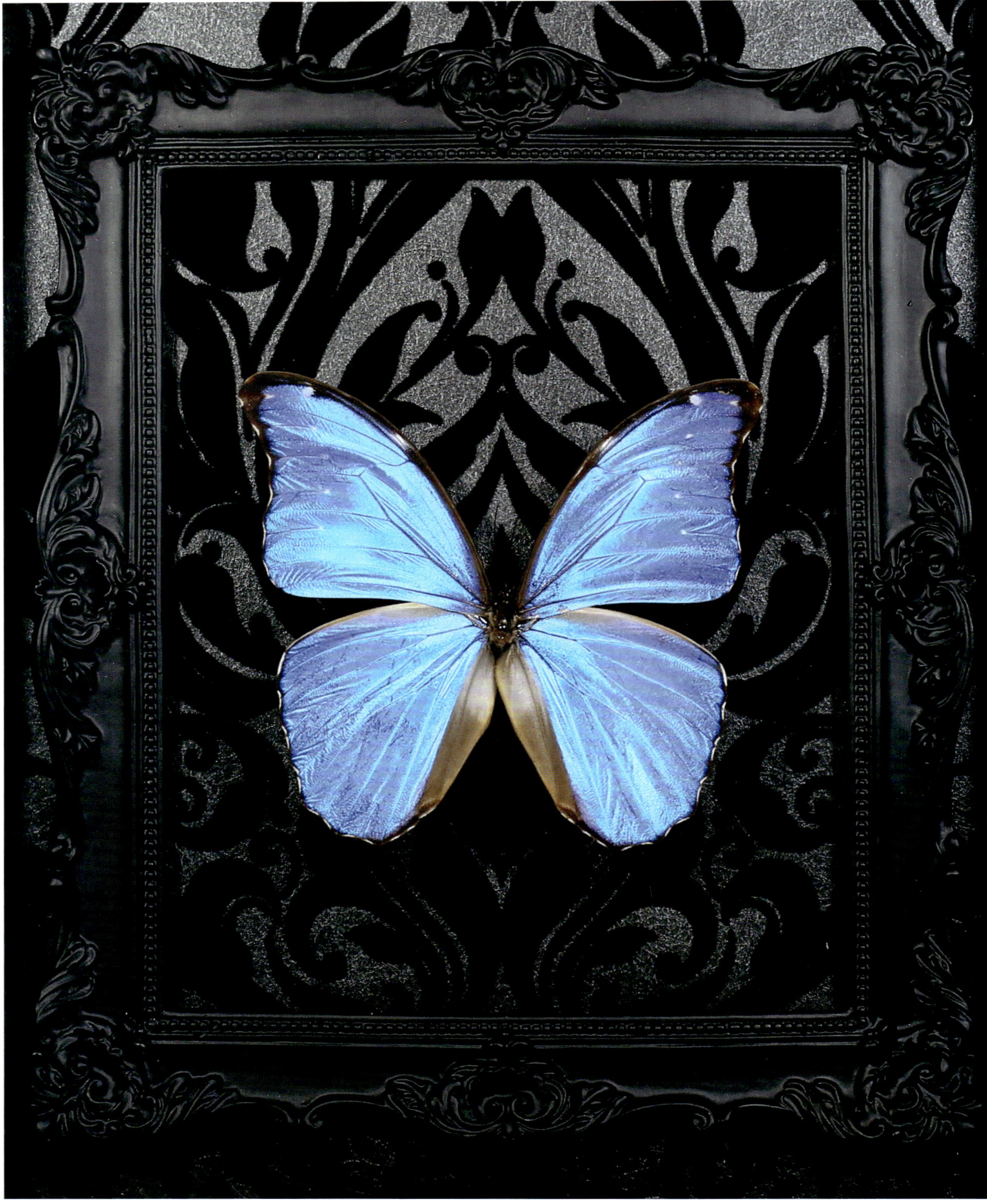

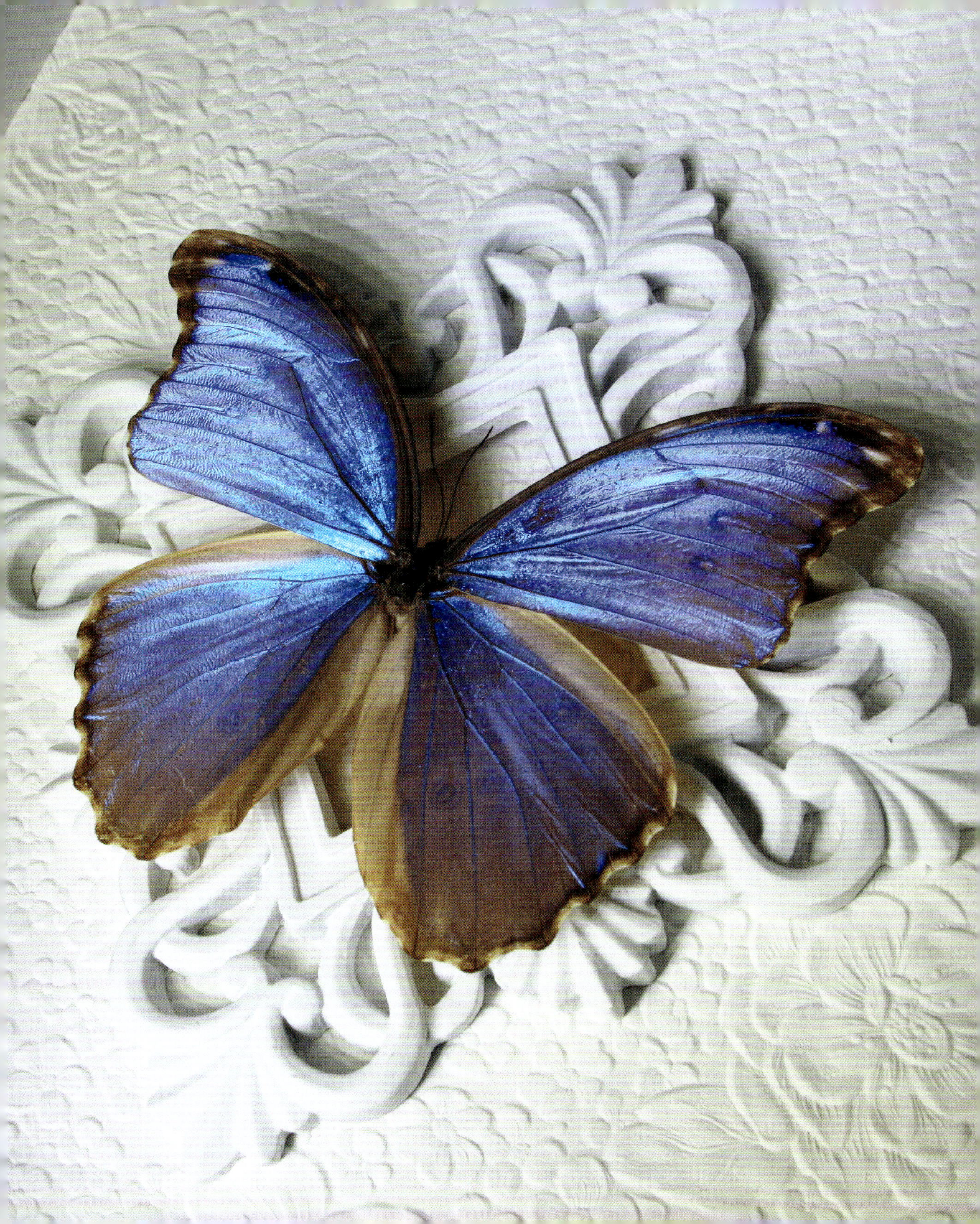

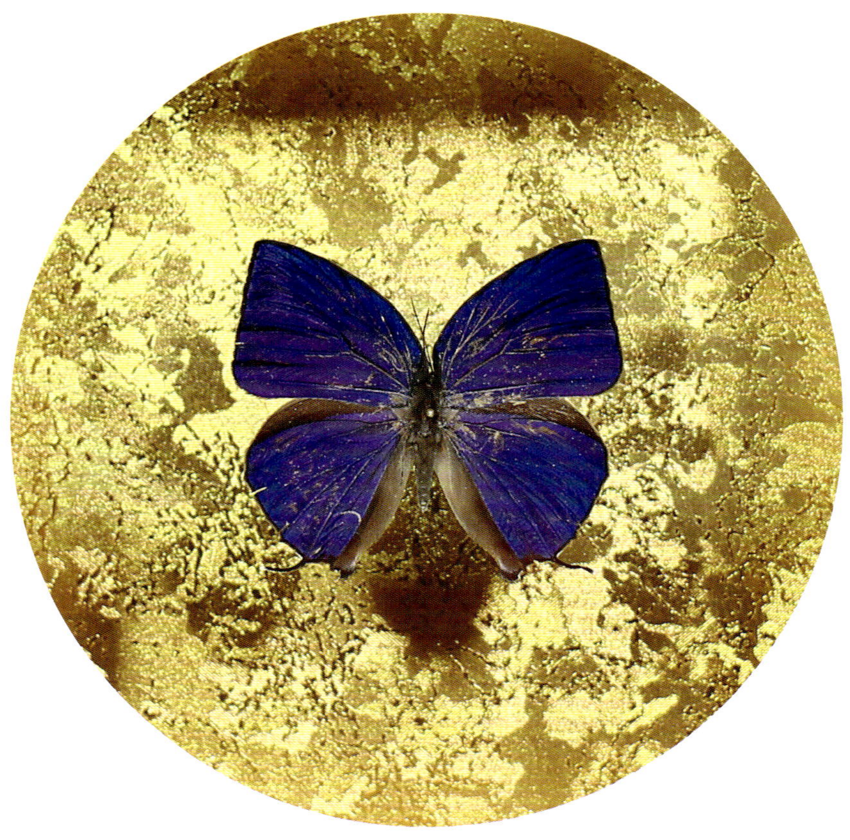

OAKBLUE BUTTERFLY

SCIENTIFIC NAME
Arhopala hercules

GEOGRAPHIC RANGE
Indonesia

The oakblue butterfly is widely recognized as the brightest solid-purple butterfly found in nature. Its vivid coloration sets it apart from other species, making it a truly impressive sight to behold. In caterpillar form, this species secretes a substance that entices ants to stay near it and help protect it from predators.

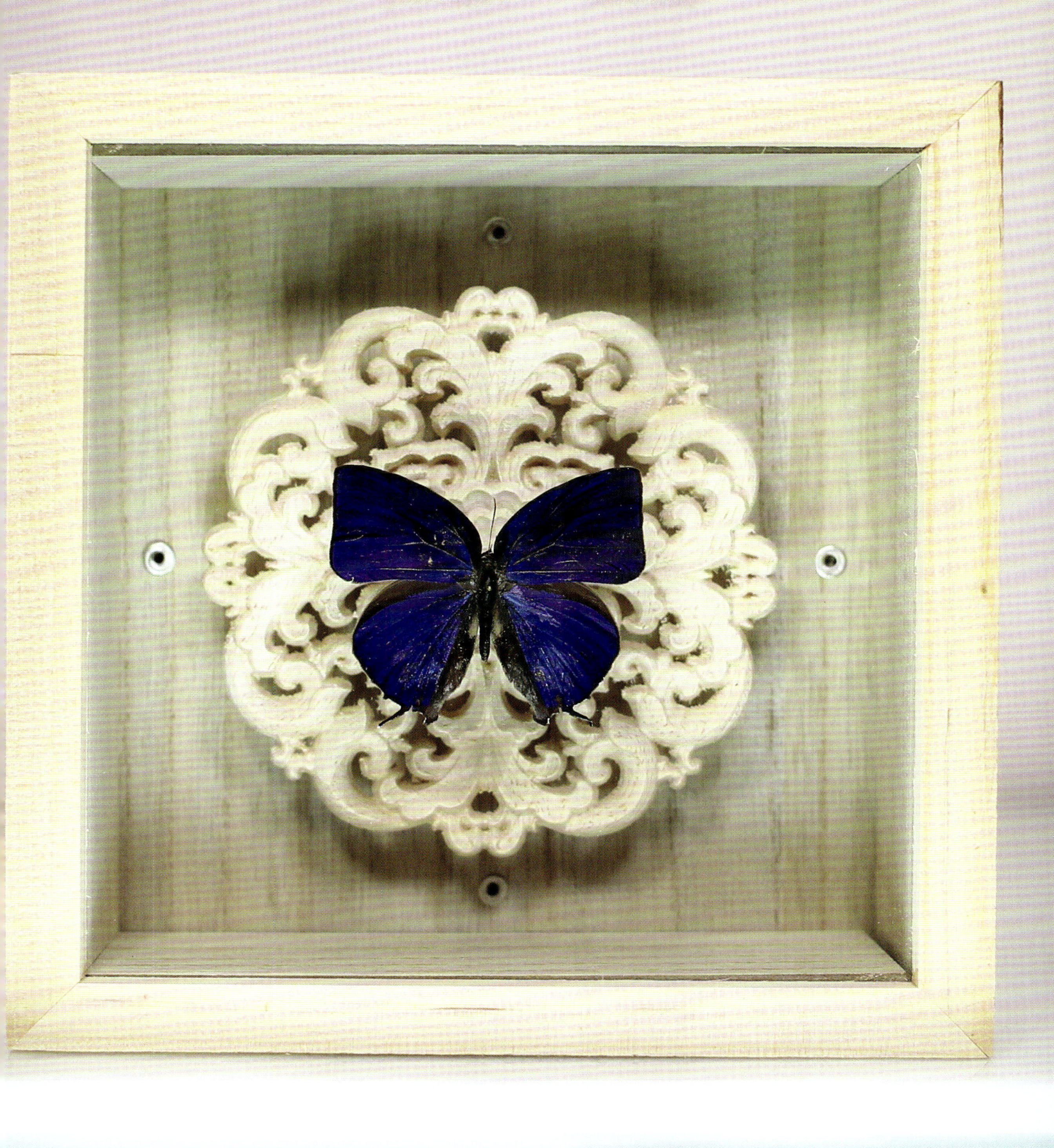

EMERALD SWALLOWTAIL

SCIENTIFIC NAME
Papilio palinurus

GEOGRAPHIC RANGE
Southeast Asia

The iridescent green sheen of this butterfly's wing pattern is created by light reflected off intricate microstructures on its wings. Visible blue and yellow reflections merge to produce the captivating green that we observe in this elegant creature.

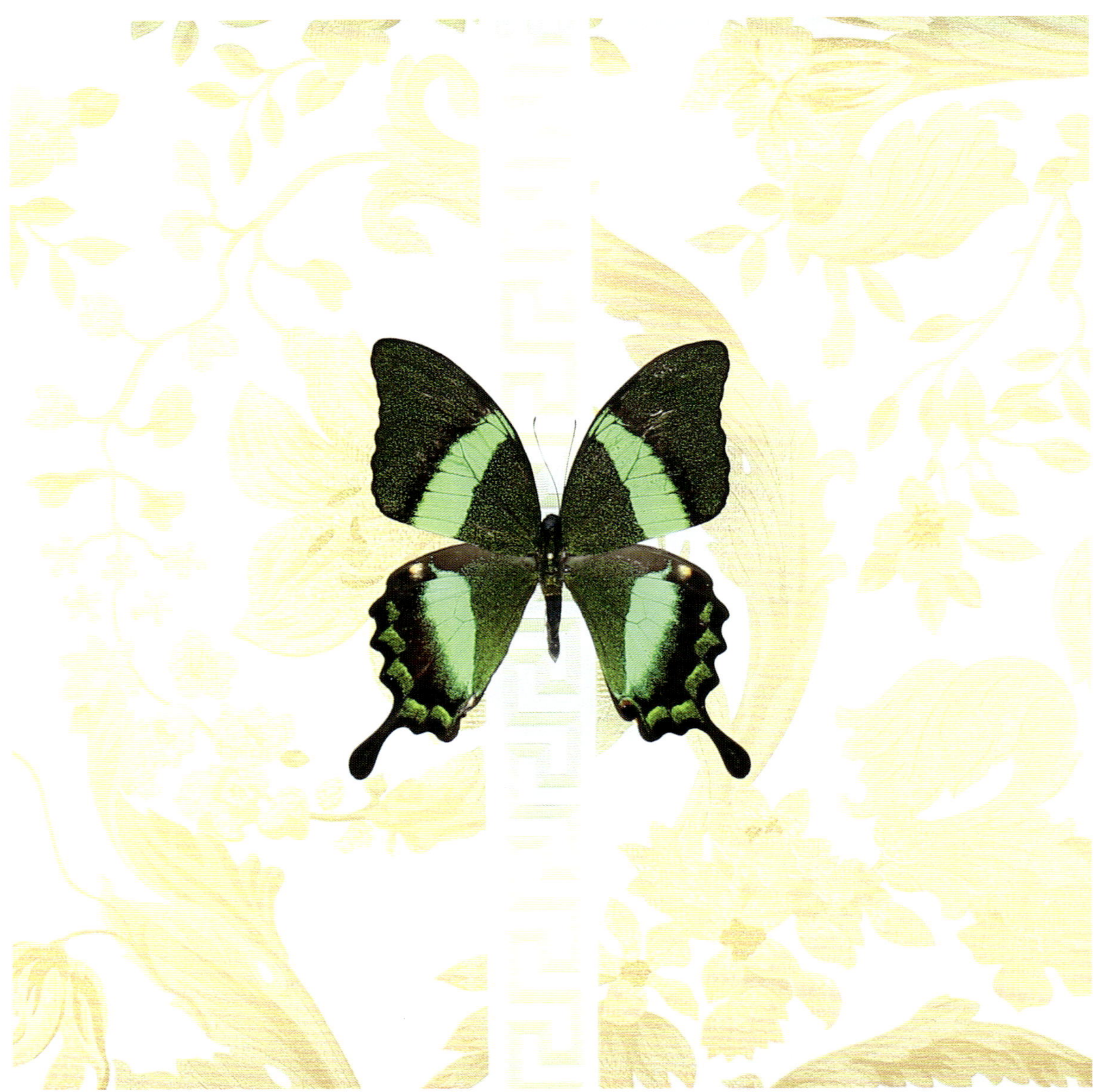

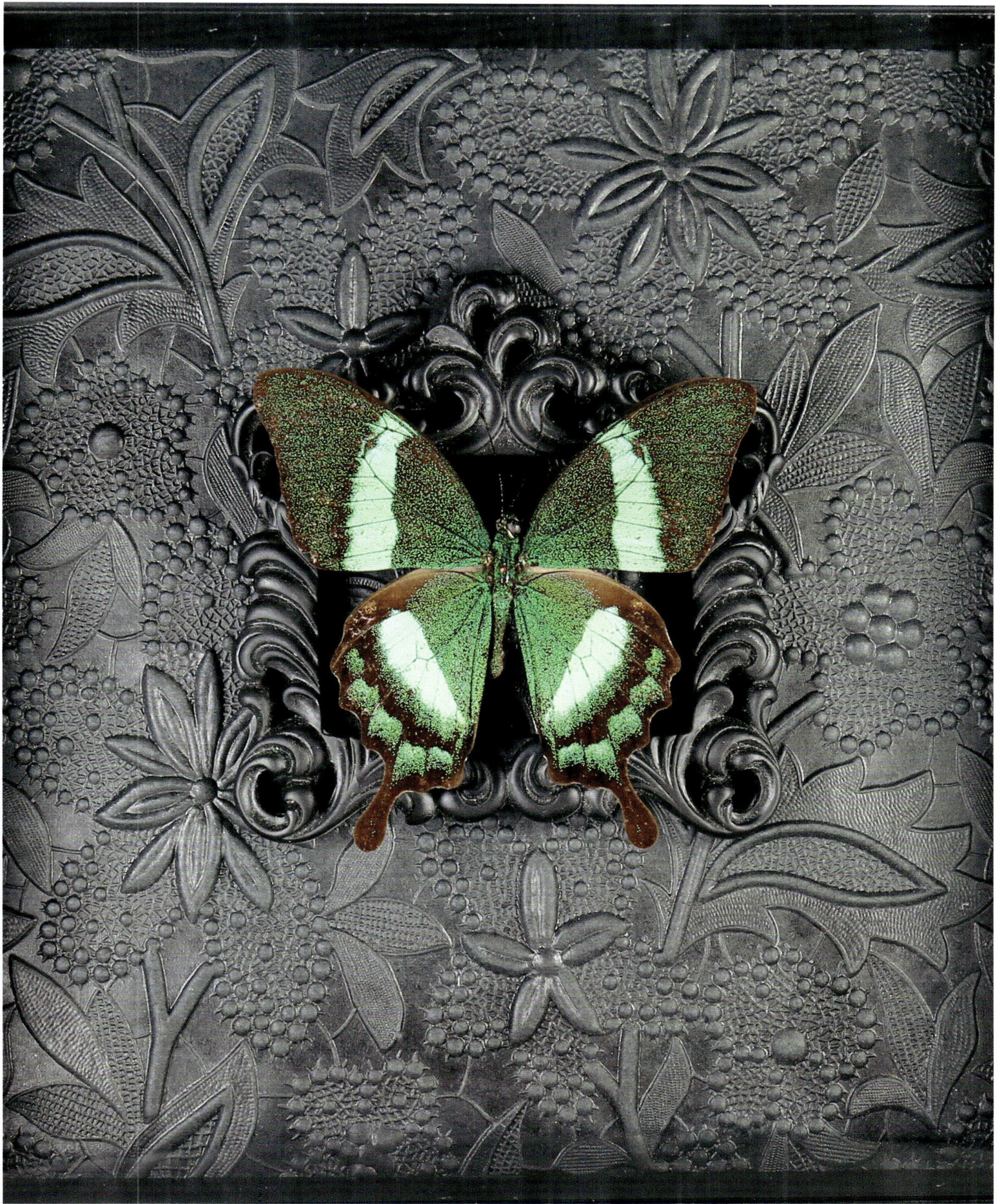

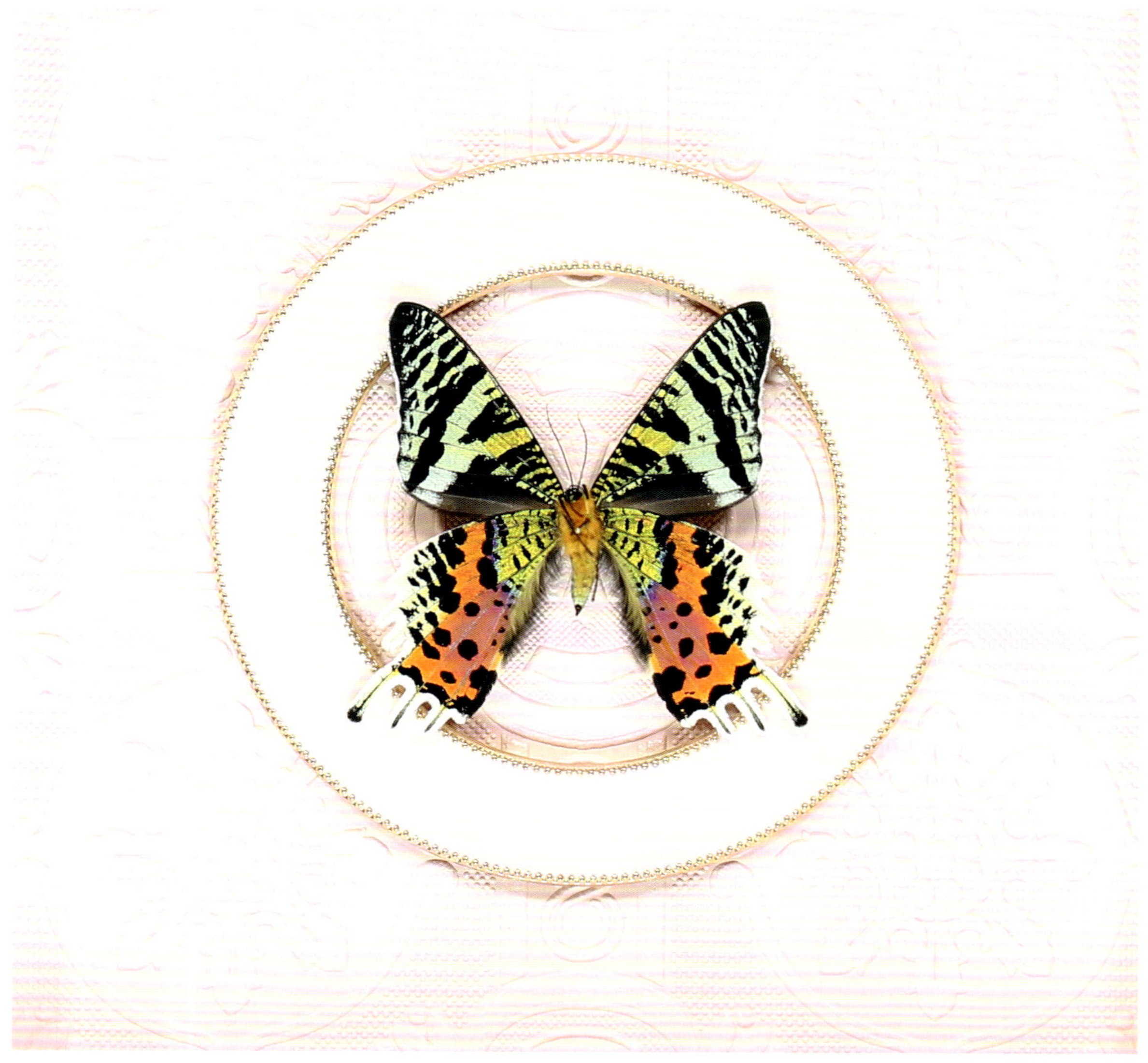

SUNSET MOTH

SCIENTIFIC NAME
Chrysiridia rhipheus

GEOGRAPHIC RANGE
Madagascar

This species is known as a specialist herbivore because its life cycle is tightly linked to the distribution of plants from the genus *Omphalea*. These plants are toxic to many other insects but a main food source for the sunset moth. It is regarded as one of the most gorgeous moths in the world, captivating many with its stunning patterns and vibrant colors.

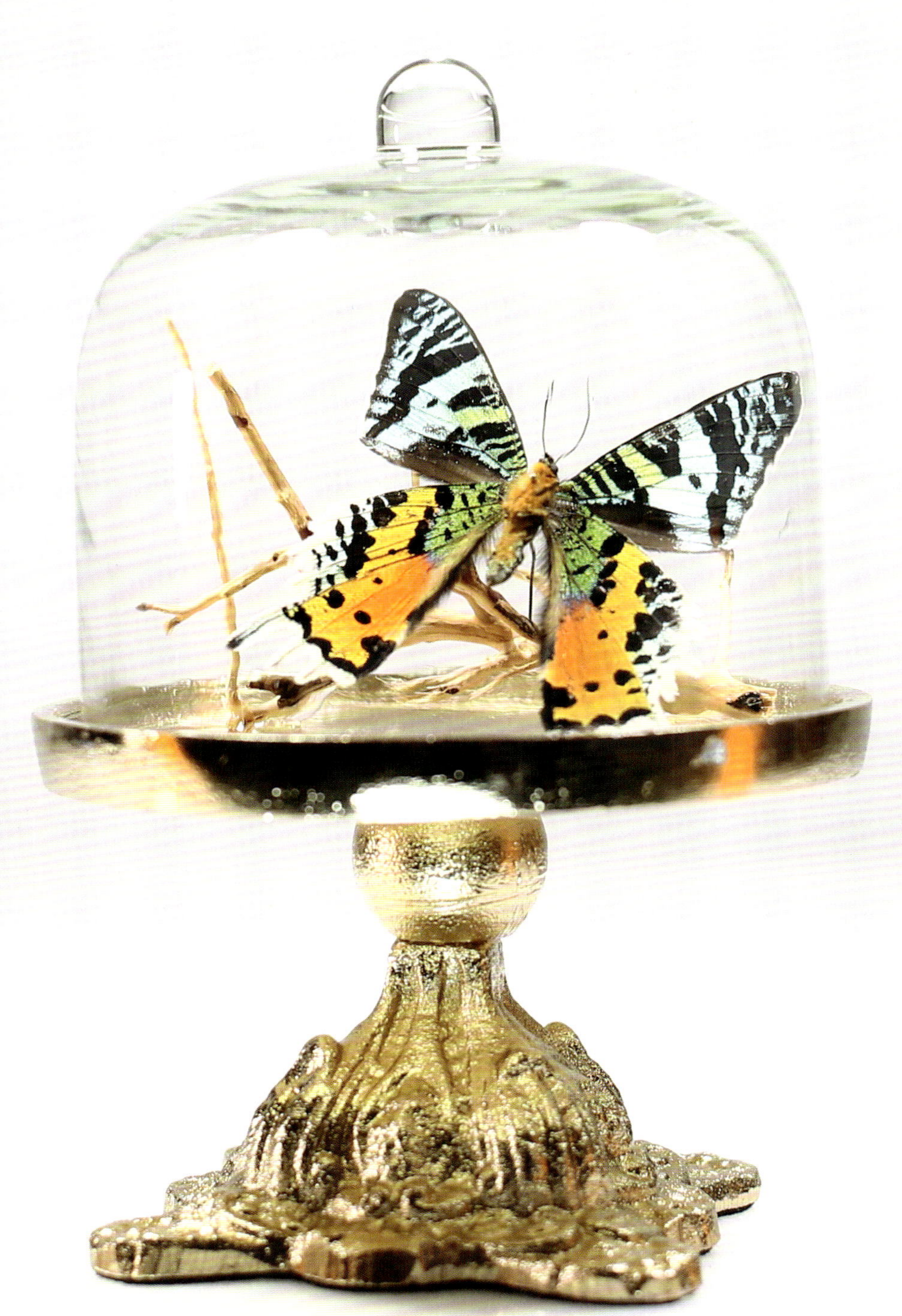

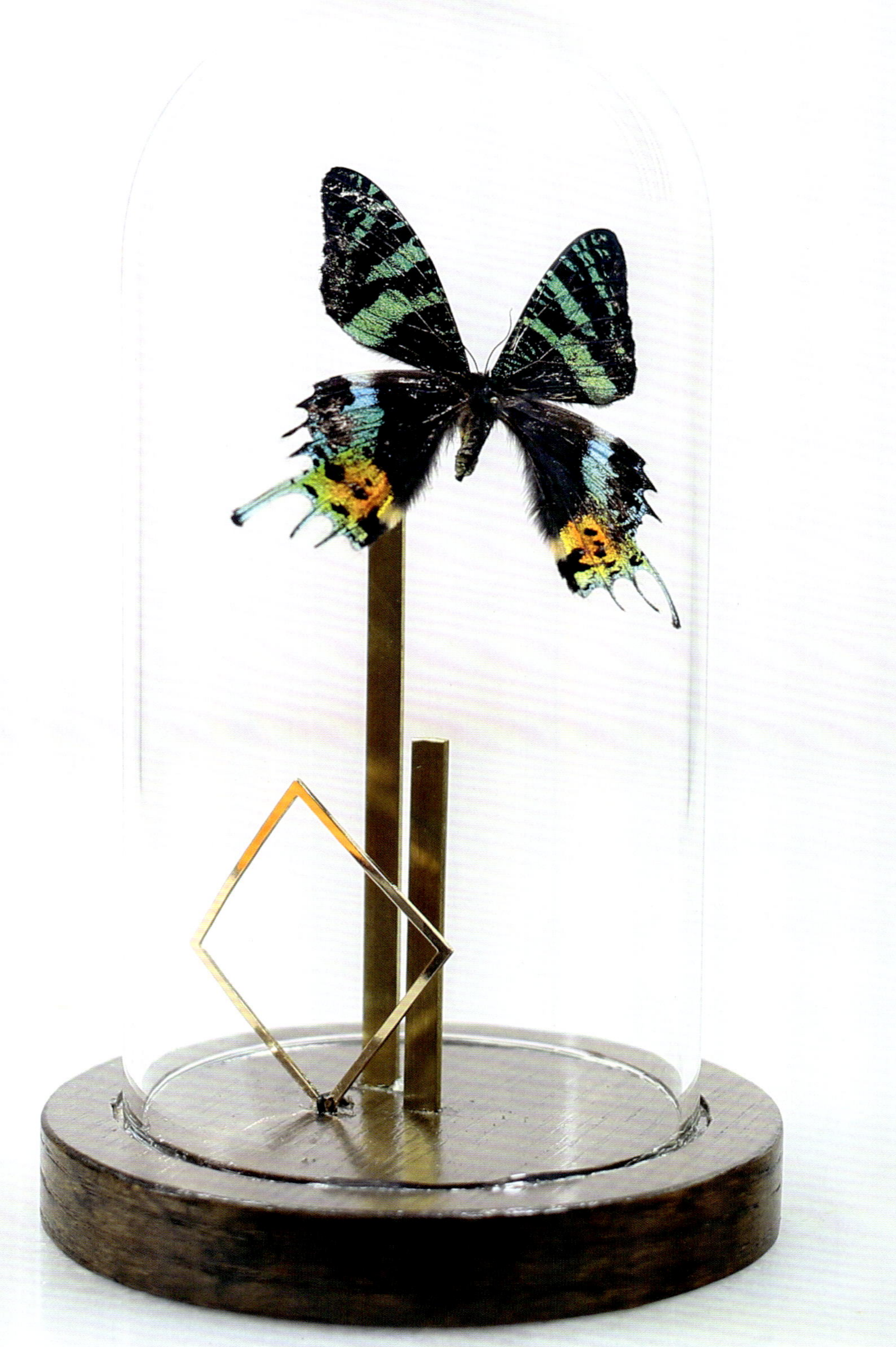

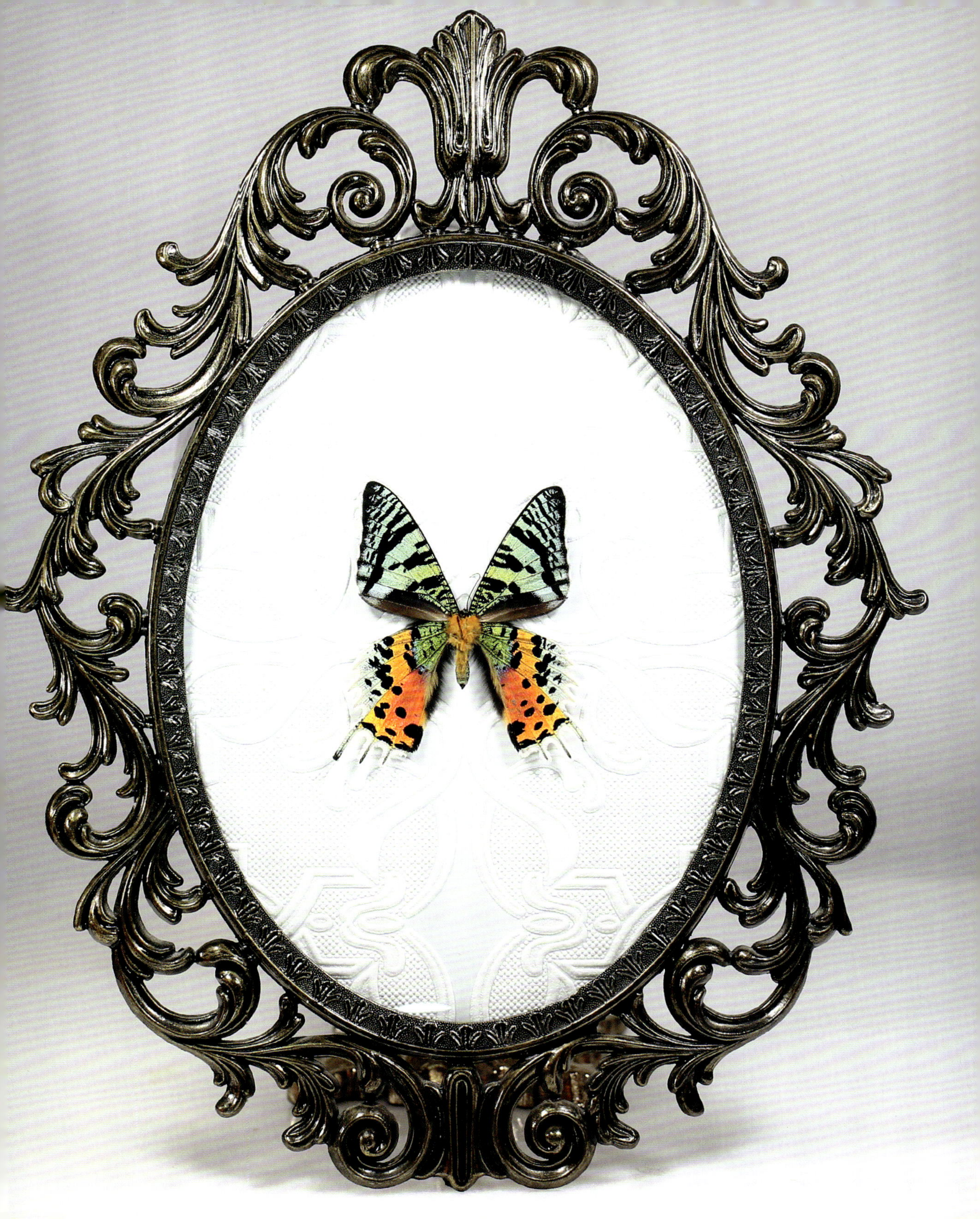

CYPRIS MORPHO

This butterfly's moniker comes from Cypris, which is another name for the enchanting Aphrodite, a Greek goddess known for her allure and irresistible charm. Males in this species are highly territorial. They often perch in the same spots repeatedly throughout the day and defend them against other males in aerial duels.

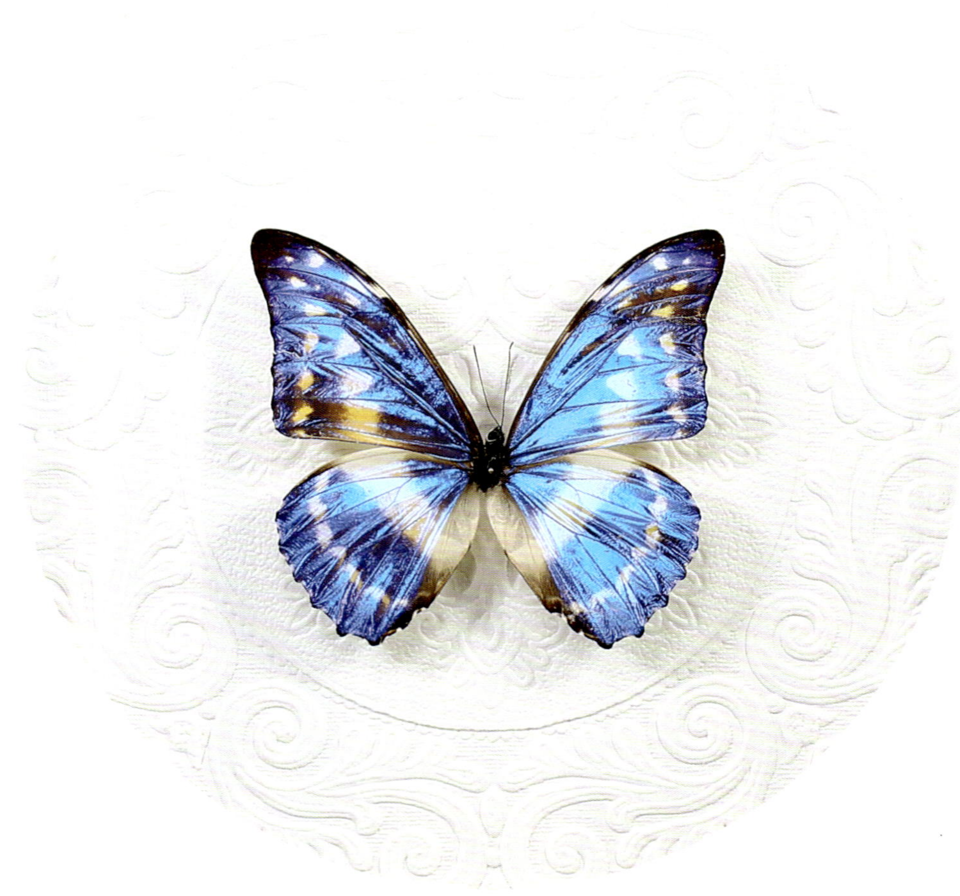

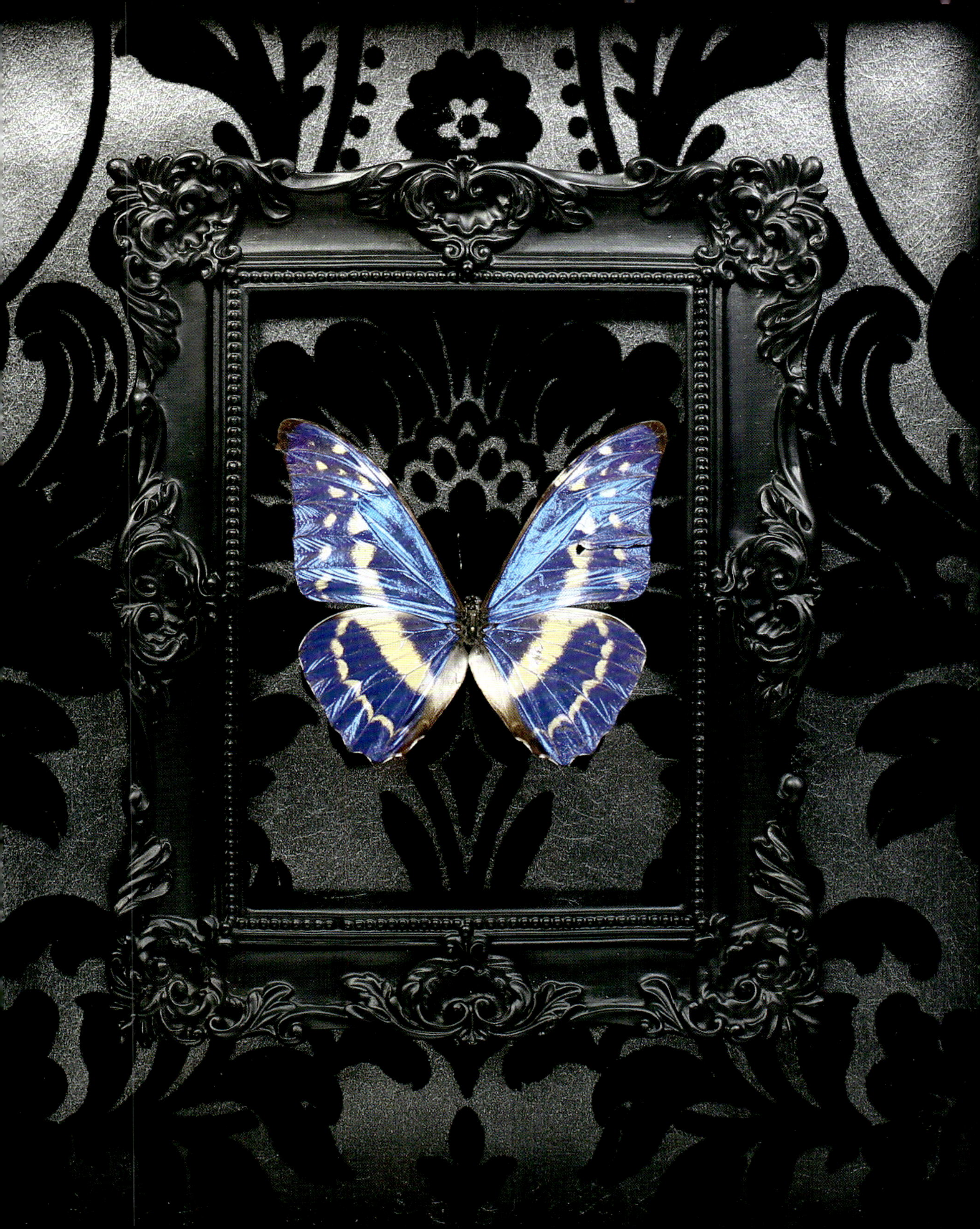

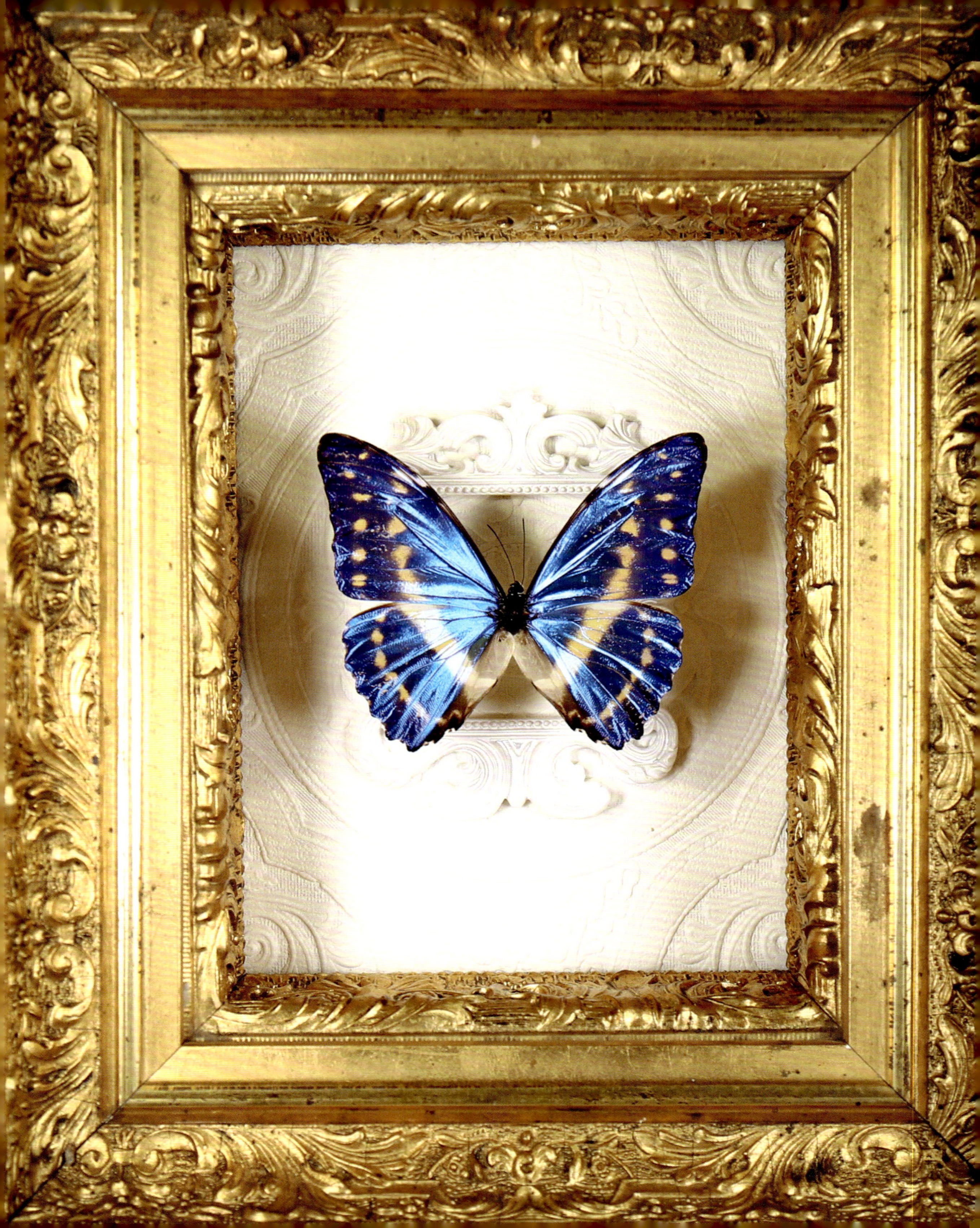

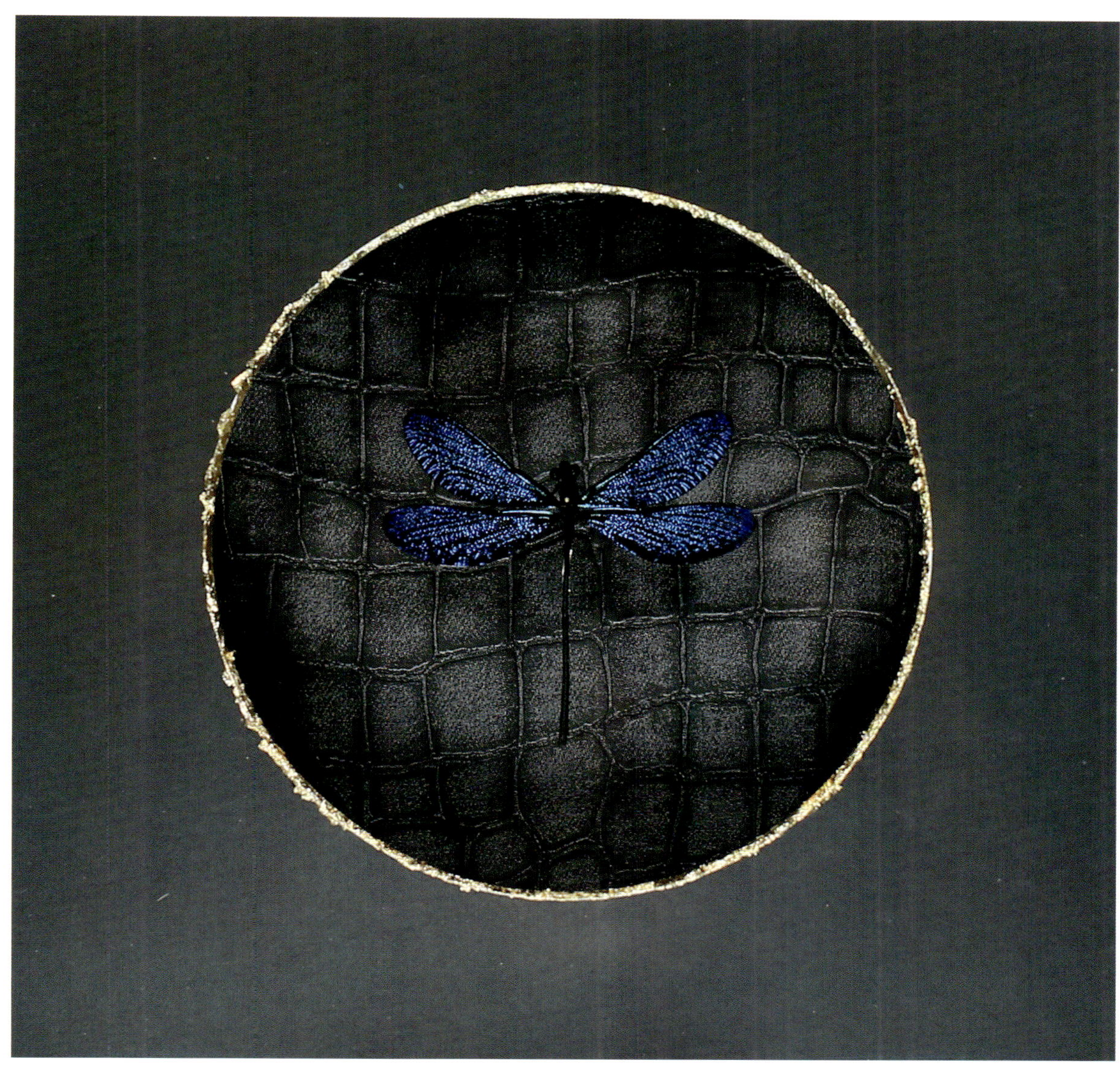

METALLIS DAMSELFLY

SCIENTIFIC NAME
Vestalis melania

GEOGRAPHIC RANGE
Philippines

Damselflies, fascinating insects belonging to the order Odonata, spend their larval stages primarily in aquatic environments, where they are highly efficient hunters. As adults, they are heliothermic, meaning that they rely on sunlight to warm their flight muscles.

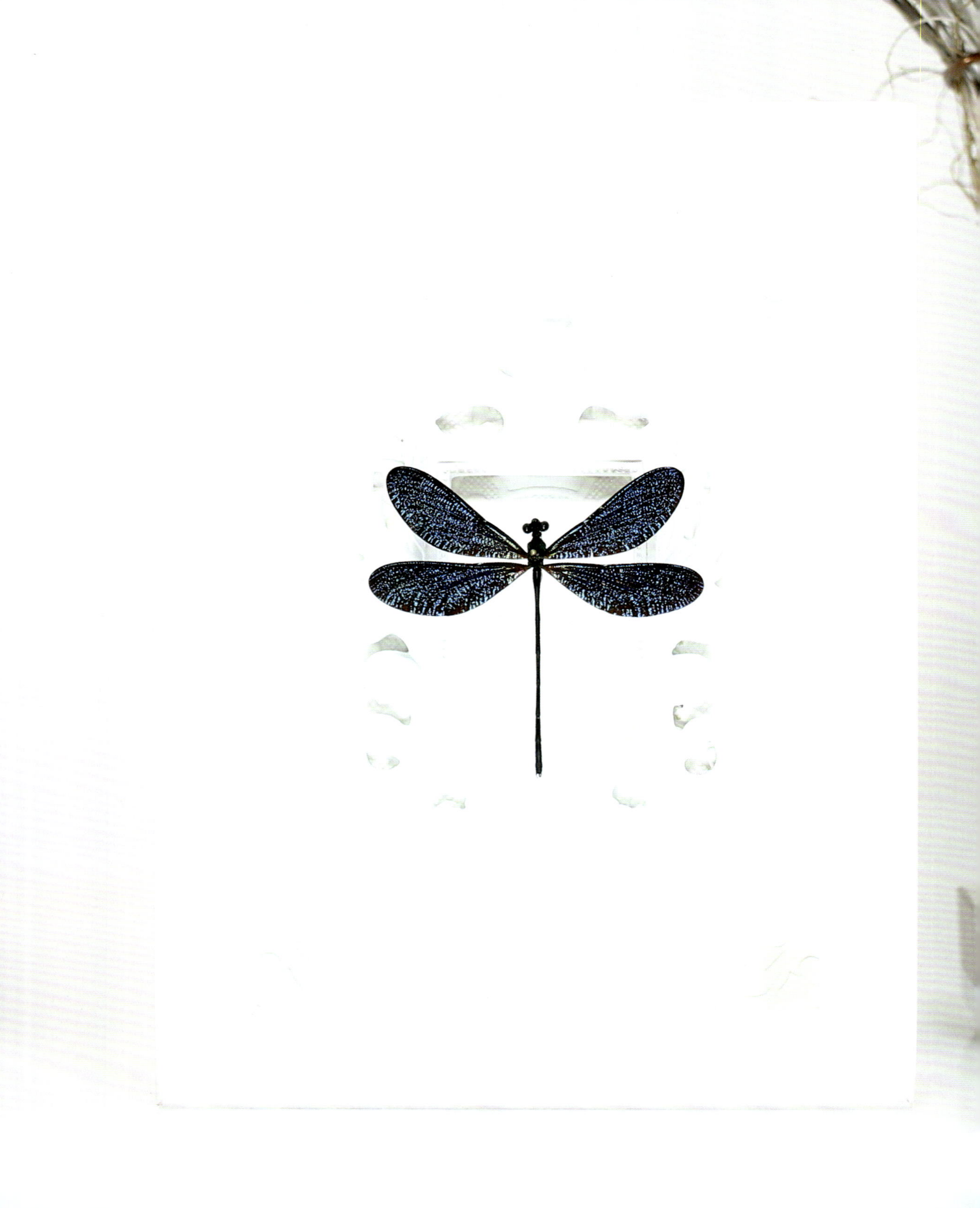

ZODIAC MOTH

SCIENTIFIC NAME
Alcides metaurus

GEOGRAPHIC RANGE
Indonesia, Papua New
Guinea, and Australia

The zodiac is a large and quite attractive moth that, in many ways, resembles a swallowtail butterfly with its vibrant colors and gorgeous patterns. It is also a day-flying moth that can be seen gracefully fluttering around in search of nectar. The caterpillars of this species undergo several distinct stages of development, showcasing a variety of vibrant colors, including a striking green adorned with a prominent black band, a deep black featuring contrasting white bands alongside a vivid red thorax, and an eye-catching red with bold black bands and orange legs.

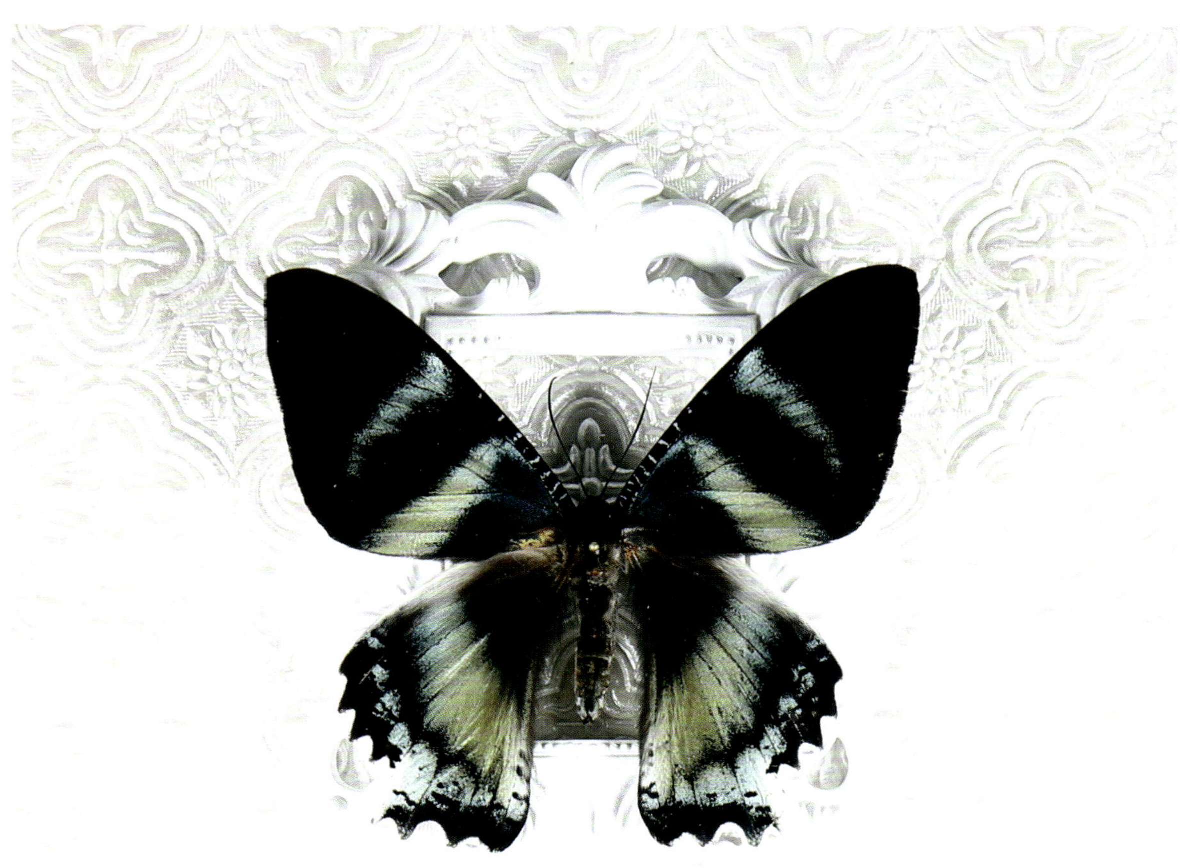

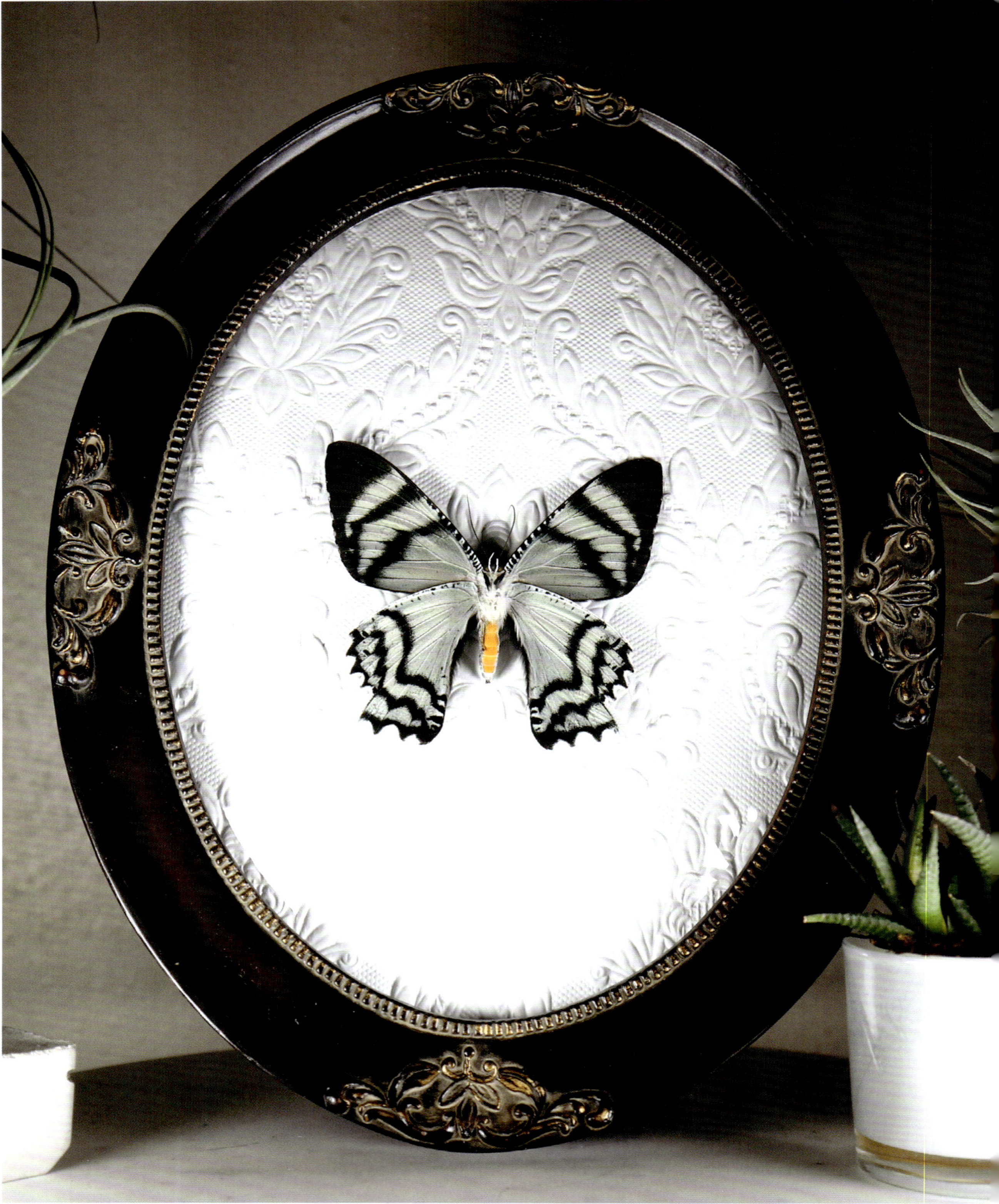

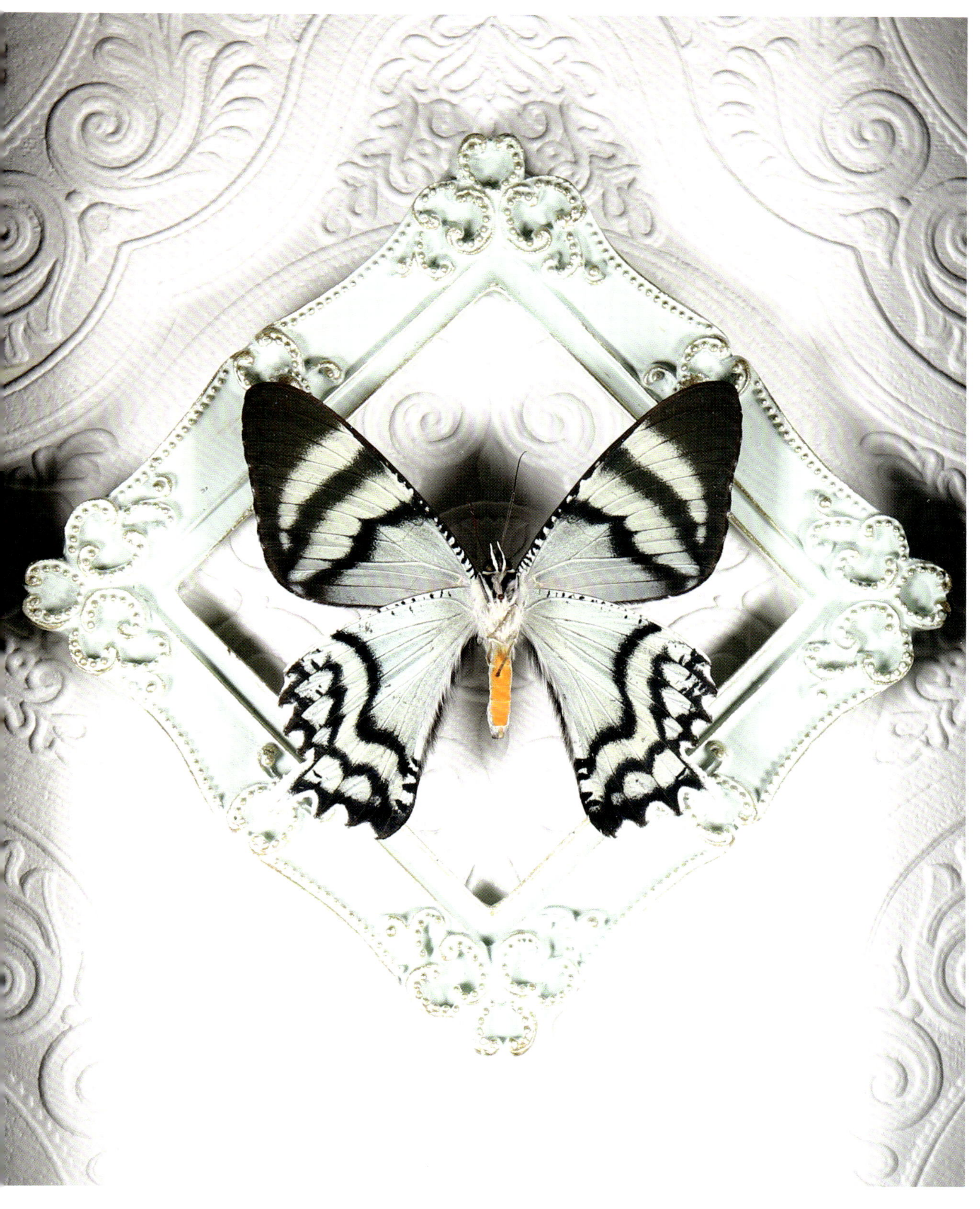

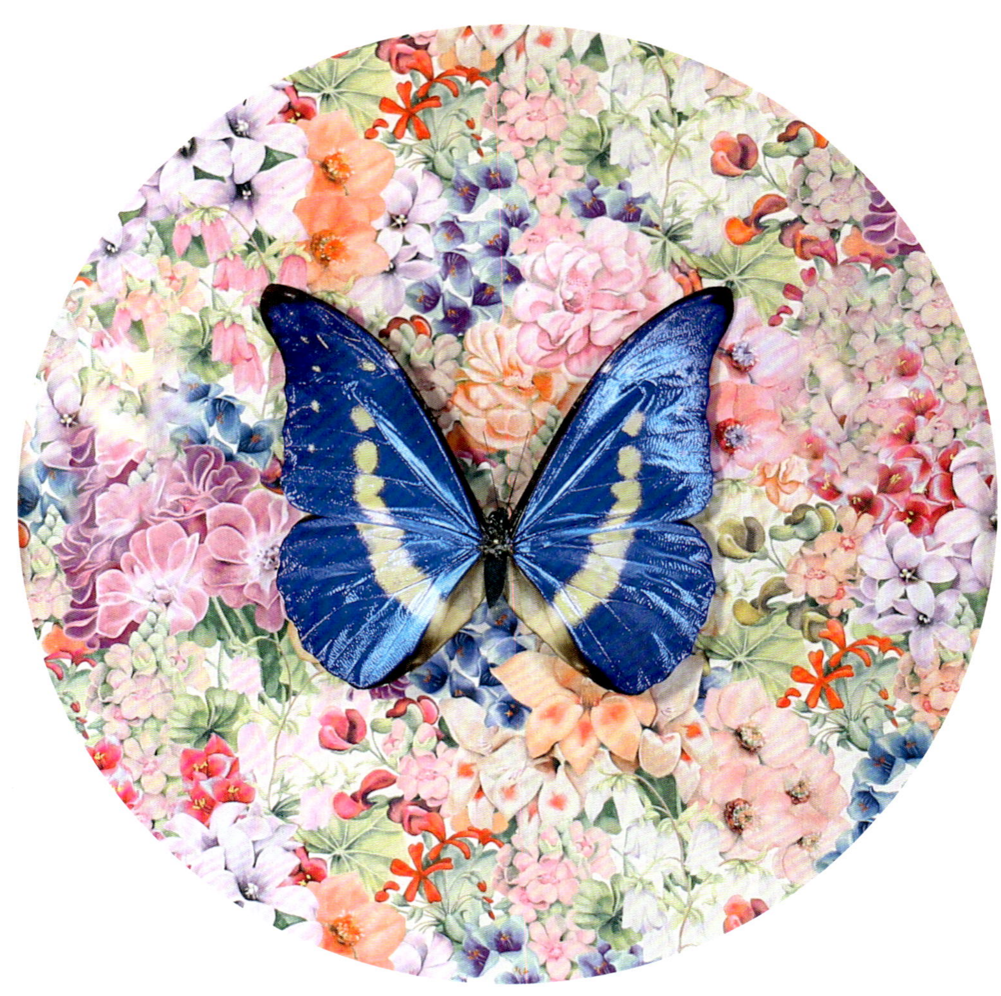

HELENA MORPHO

SCIENTIFIC NAME
Morpho helena

GEOGRAPHIC RANGE
Peru

The blue wings of morpho butterflies shimmer and catch the light so beautifully that pilots flying over the vast expanse of the Amazon rainforest have reported spotting such creatures (likely even Helenas) fluttering above the lush greenery.

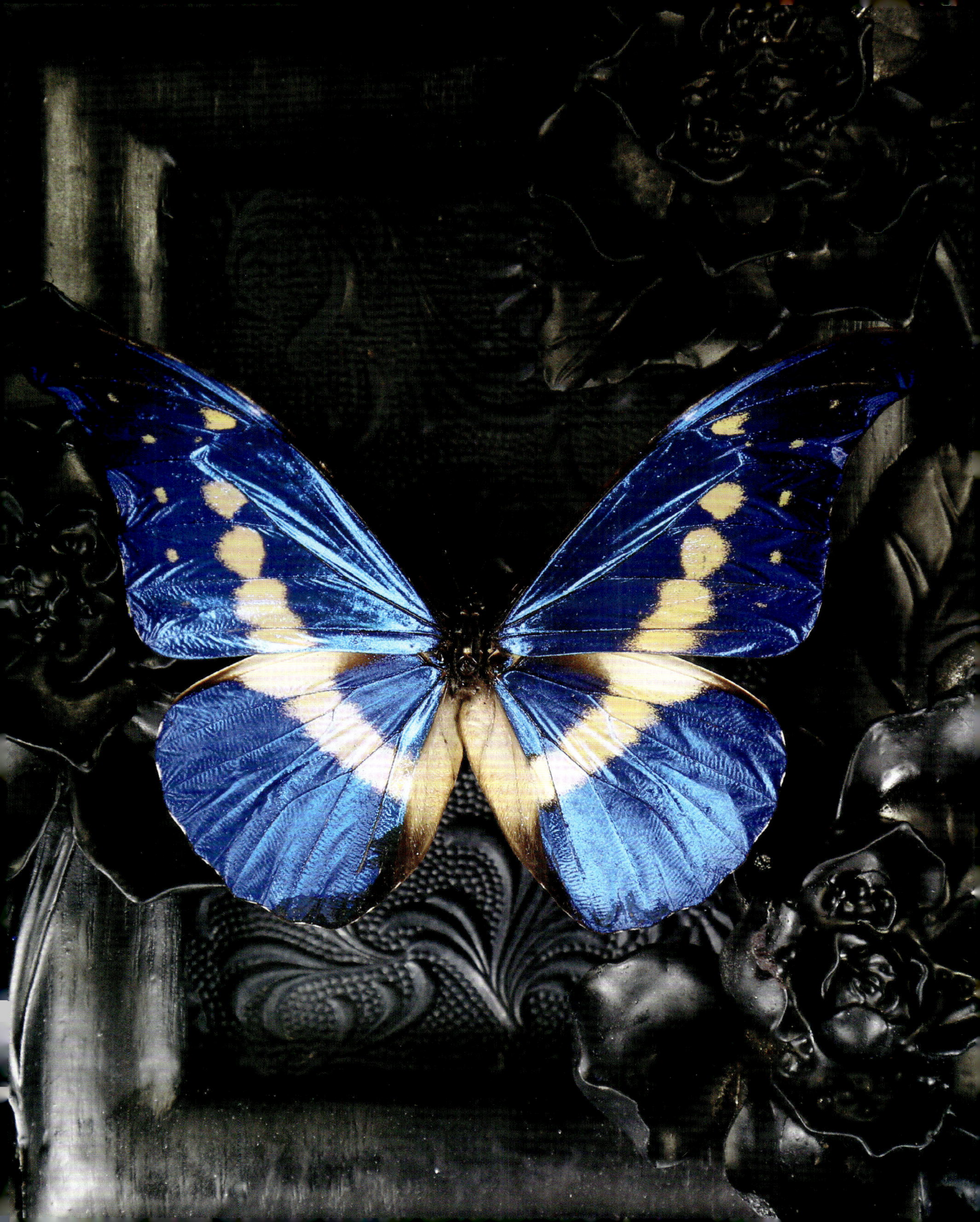

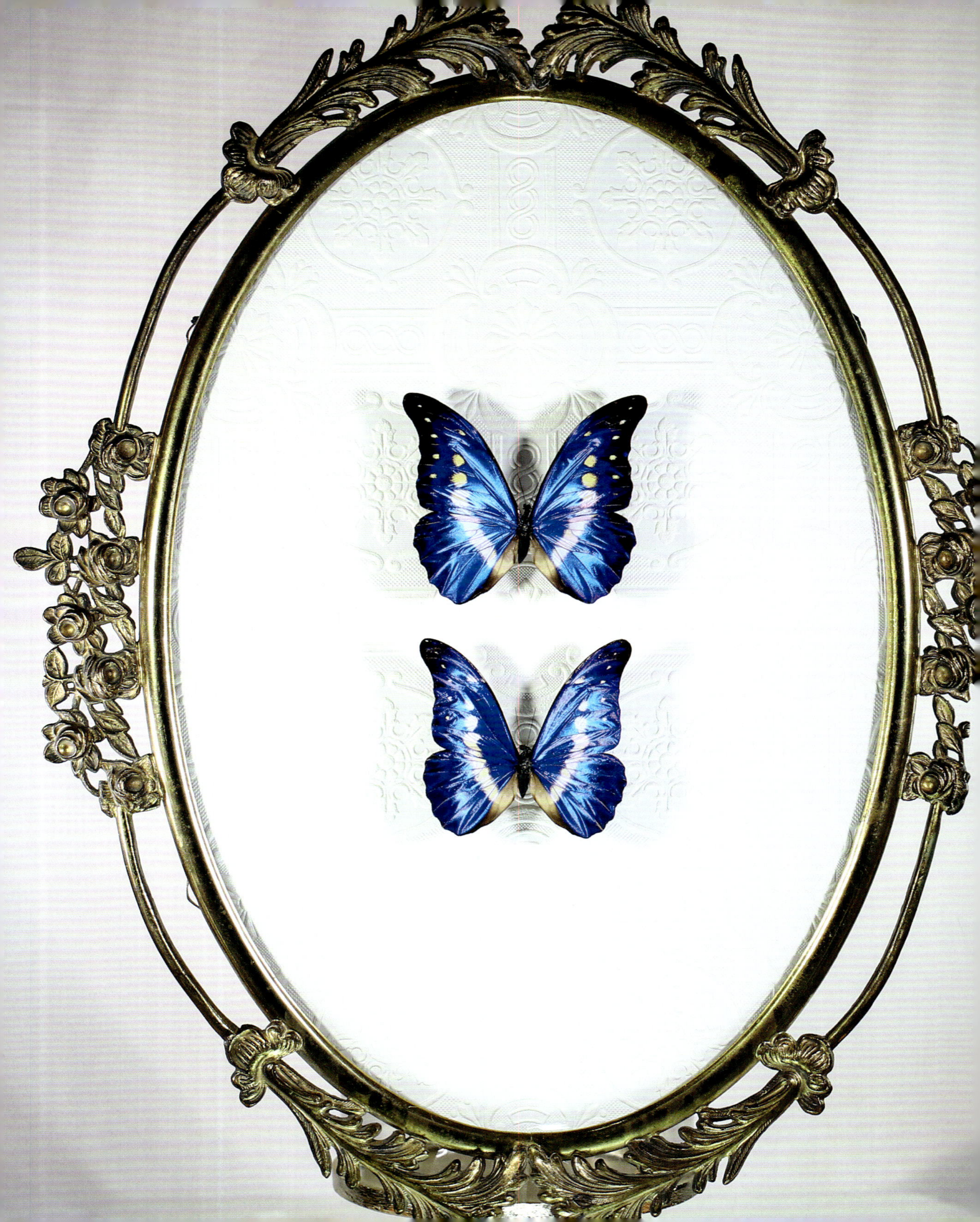

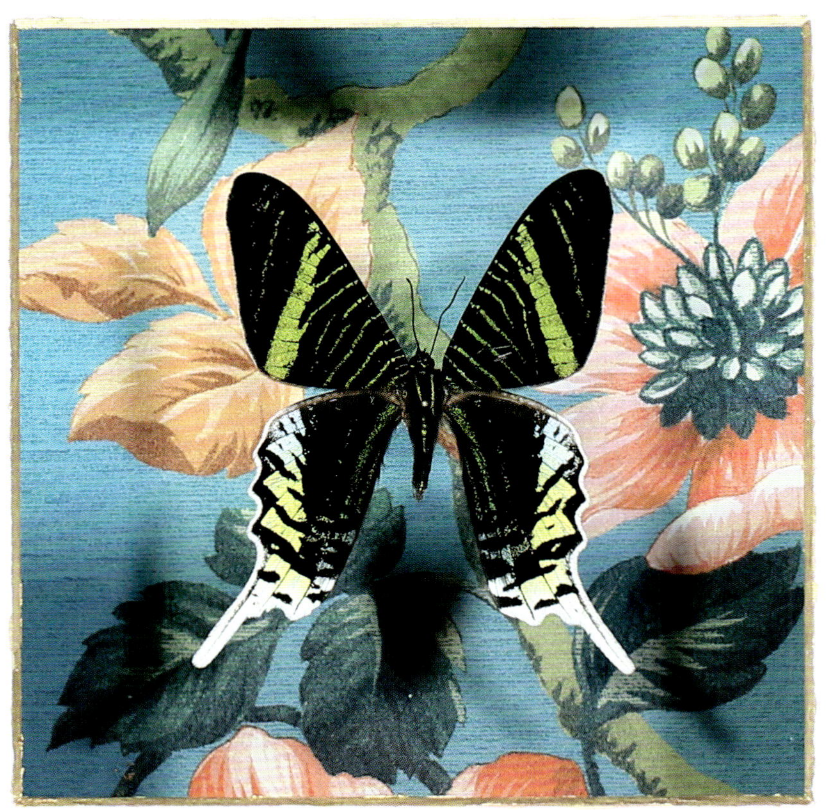

GREEN-BANDED URANIA MOTH

SCIENTIFIC NAME
Urania leilus

GEOGRAPHIC RANGE
South America

Most moths have ears (tympanic organs) to detect predatory bats at night. But because this moth is strictly active during the day and rests hidden at night, it's thought to have lost these auditory structures over evolutionary time.

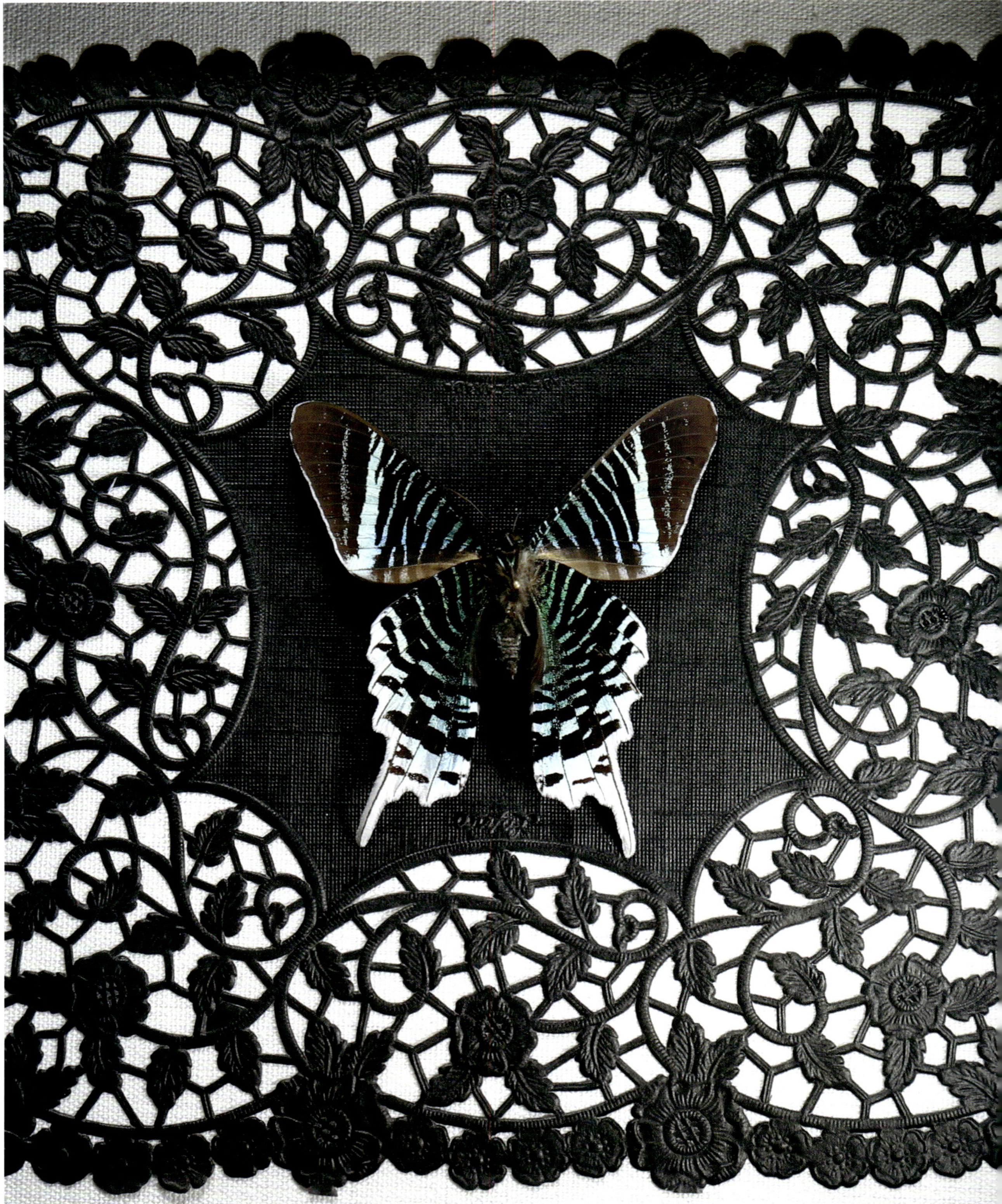

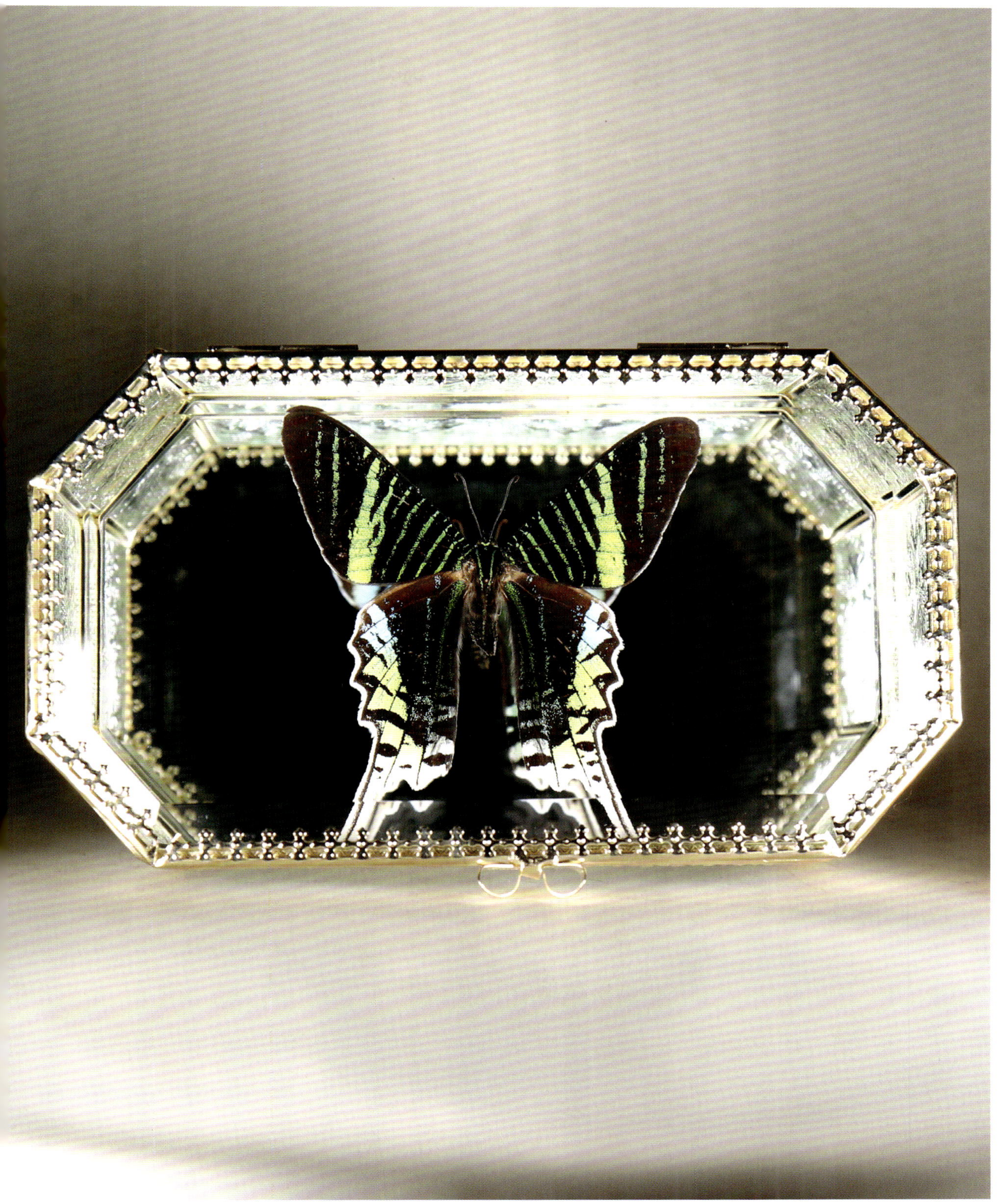

PEARL MORPHO

SCIENTIFIC NAME
Morpho sulkowskyi

GEOGRAPHIC RANGE
Colombia, Ecuador,
Peru, and Bolivia

This butterfly may appear white, but a closer look reveals the special pearlescent glow that gives it its name. Our eyes perceive this kind of shine because the pearl morpho's wings have both fluorescent pigmentation and a unique nanostructure. This butterfly is found in the lush tropical cloud forests of the Andes, where it thrives at elevations as high as 11,483 feet (3,500 meters) above sea level.

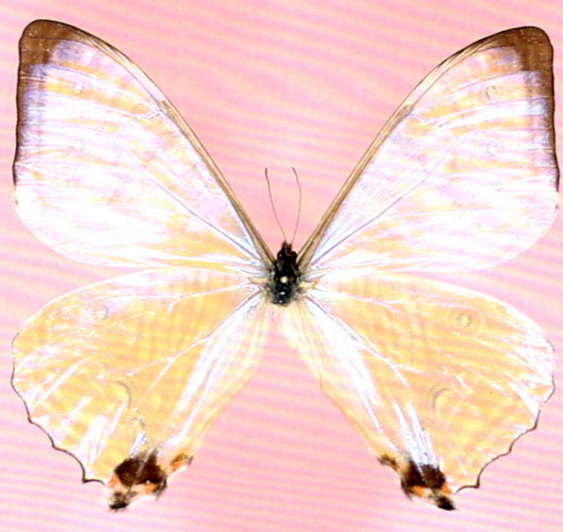

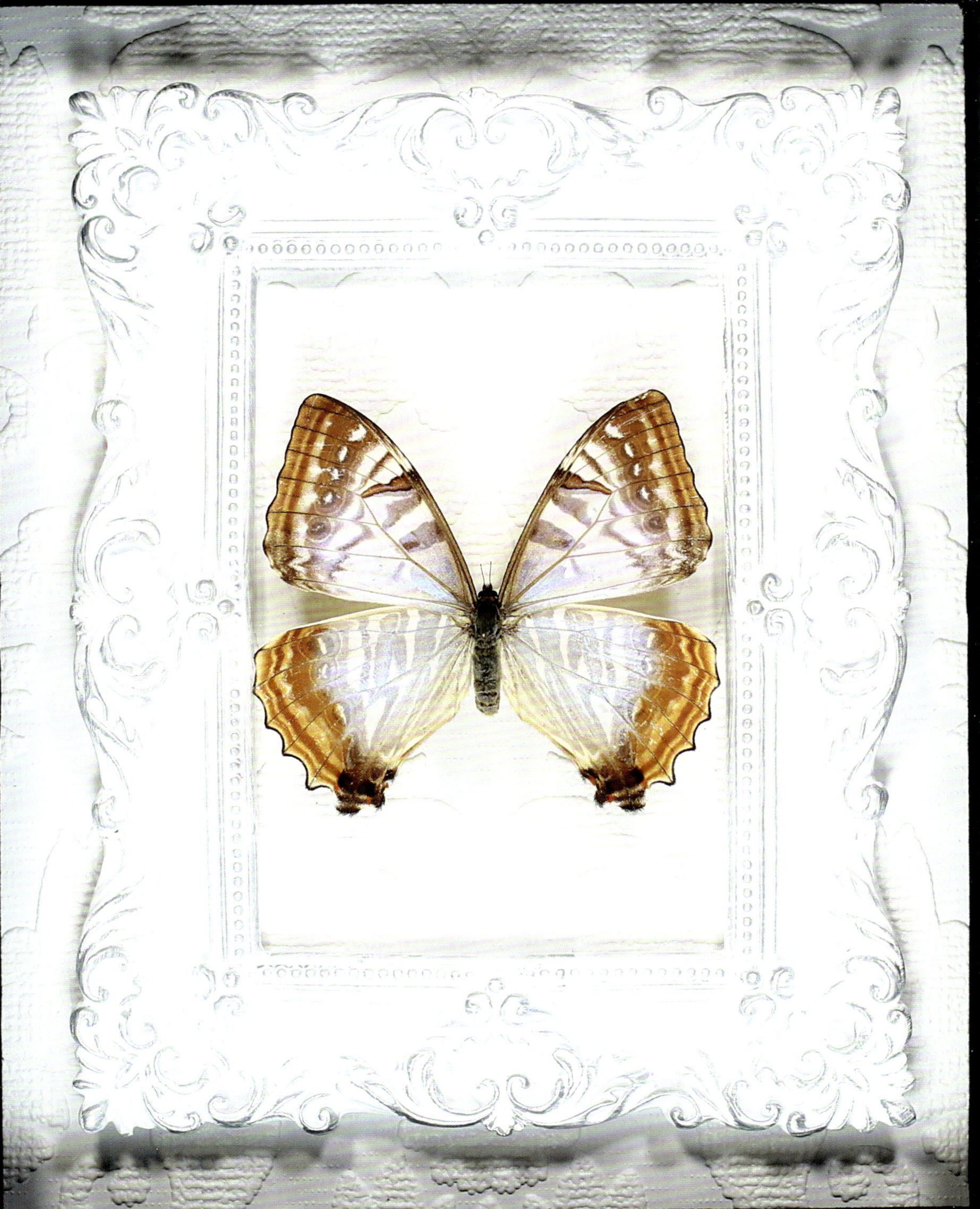

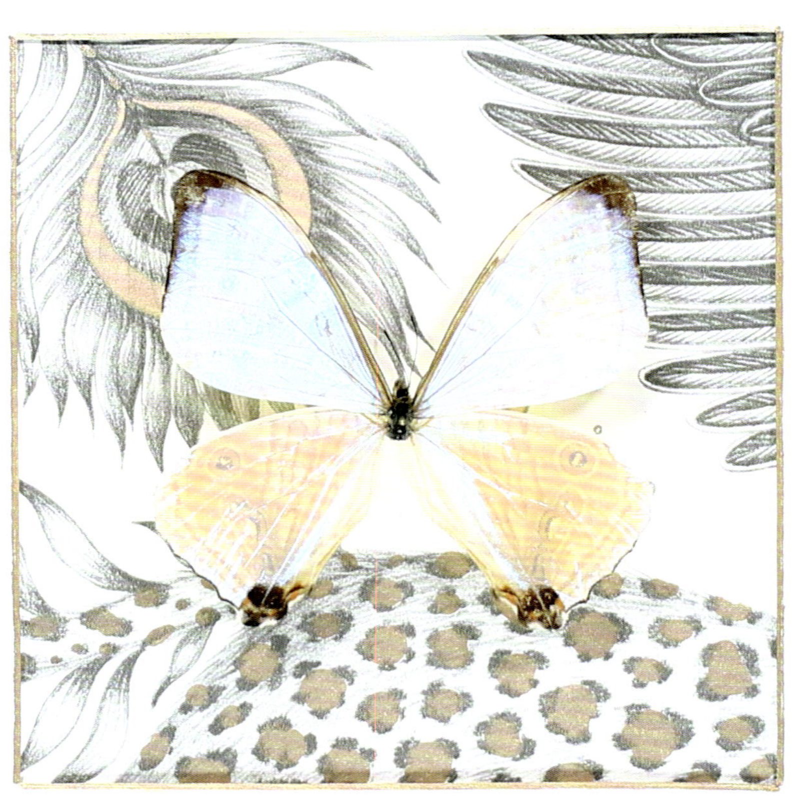

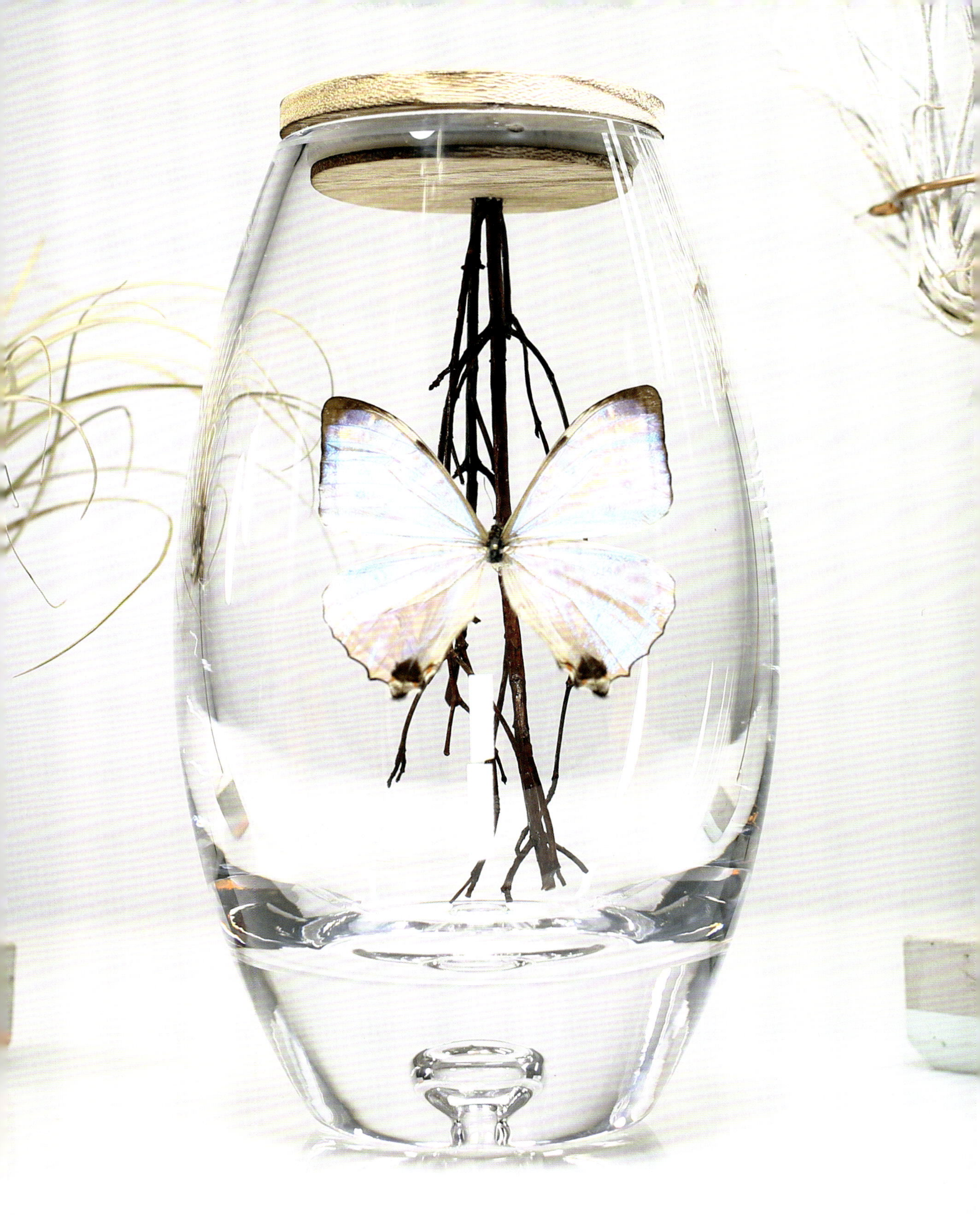

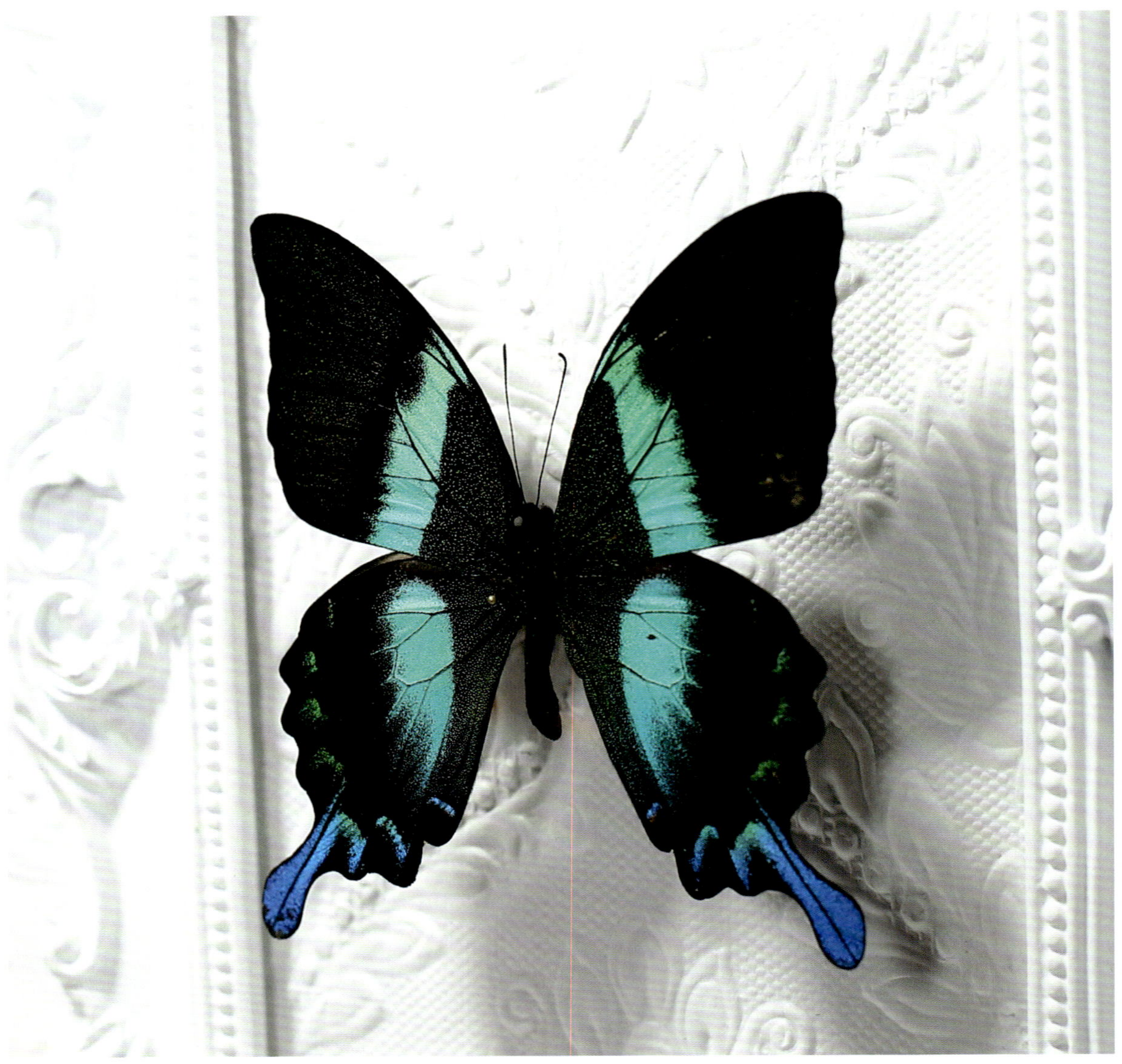

PEACOCK SWALLOWTAIL

SCIENTIFIC NAME
Papilio blumei

GEOGRAPHIC RANGE
Indonesia

During their early development stages, peacock swallowtail larvae are brown and white, resembling fresh bird droppings. This form of visual deception, called masquerade mimicry, actively repels predators by exploiting their aversion to unappetizing objects.

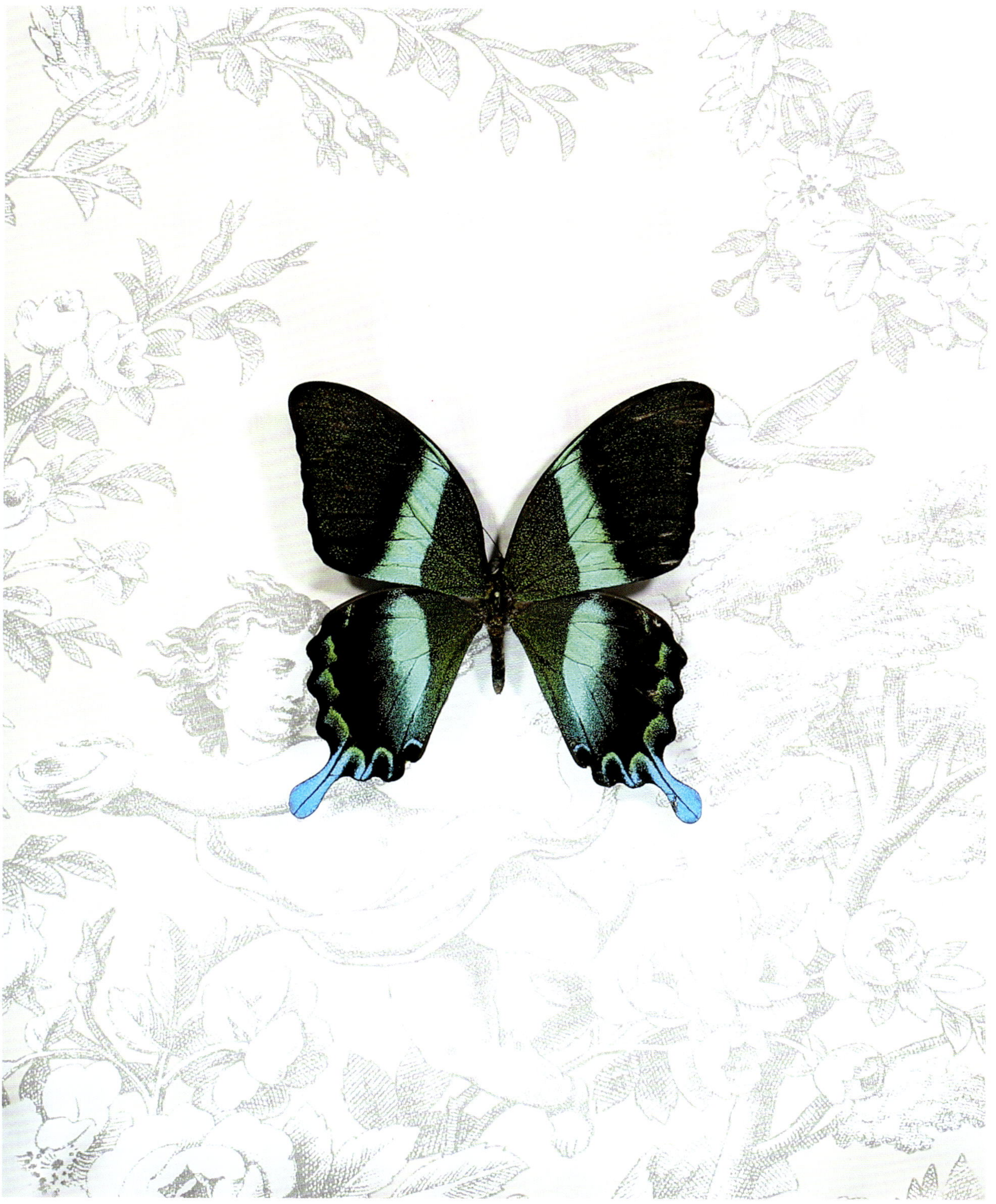

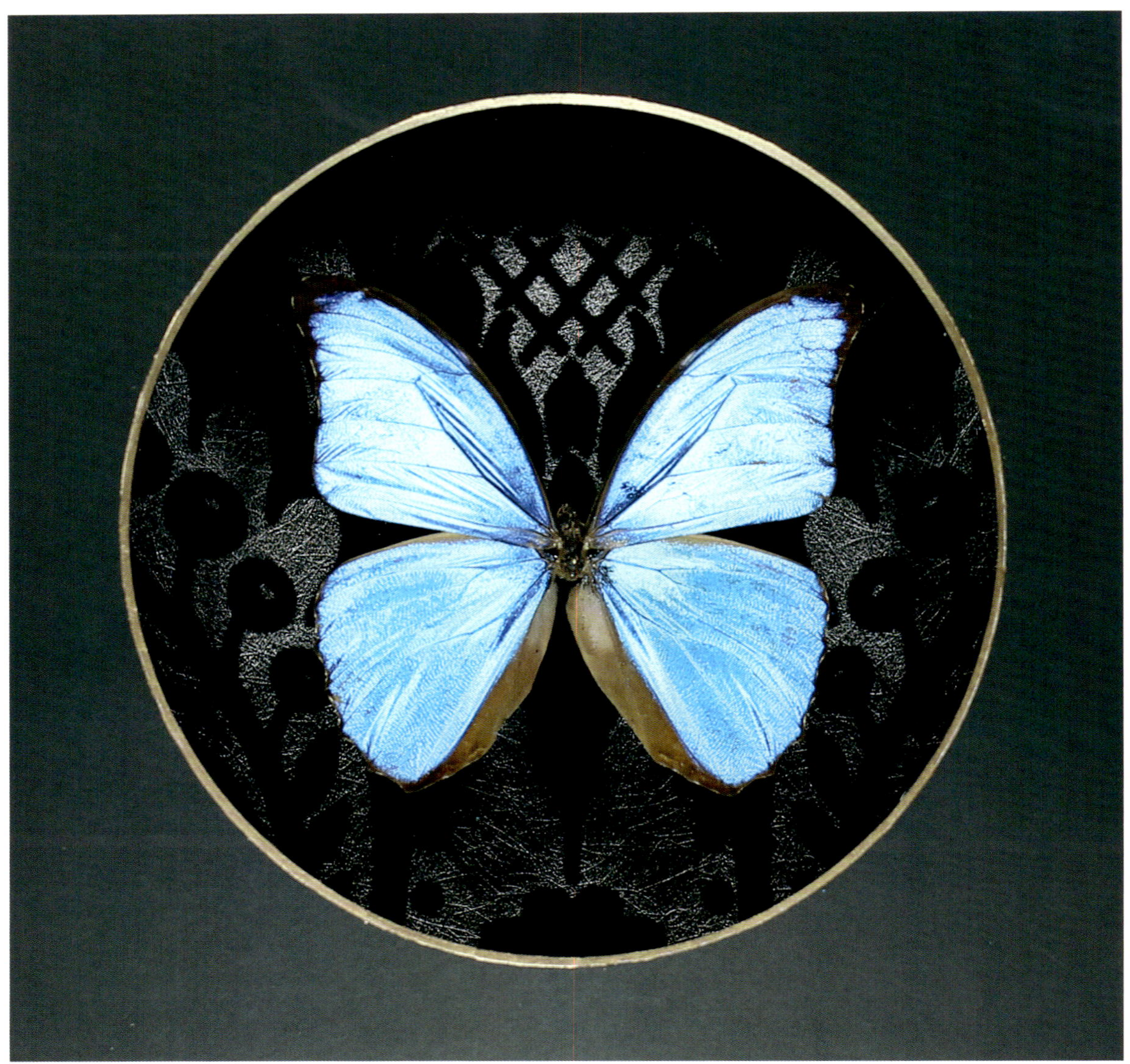

BLUE MORPHO BUTTERFLY

SCIENTIFIC NAME
Morpho menelaus

GEOGRAPHIC RANGE
Central America and
South America

The flight of the blue morpho butterfly is slow and looping, which might make it seem like an easy target—but this is actually a deceptive strategy. When pursued, it abruptly dives into the forest understory and closes its wings, vanishing into the shadows, thanks to its dull brown underside.

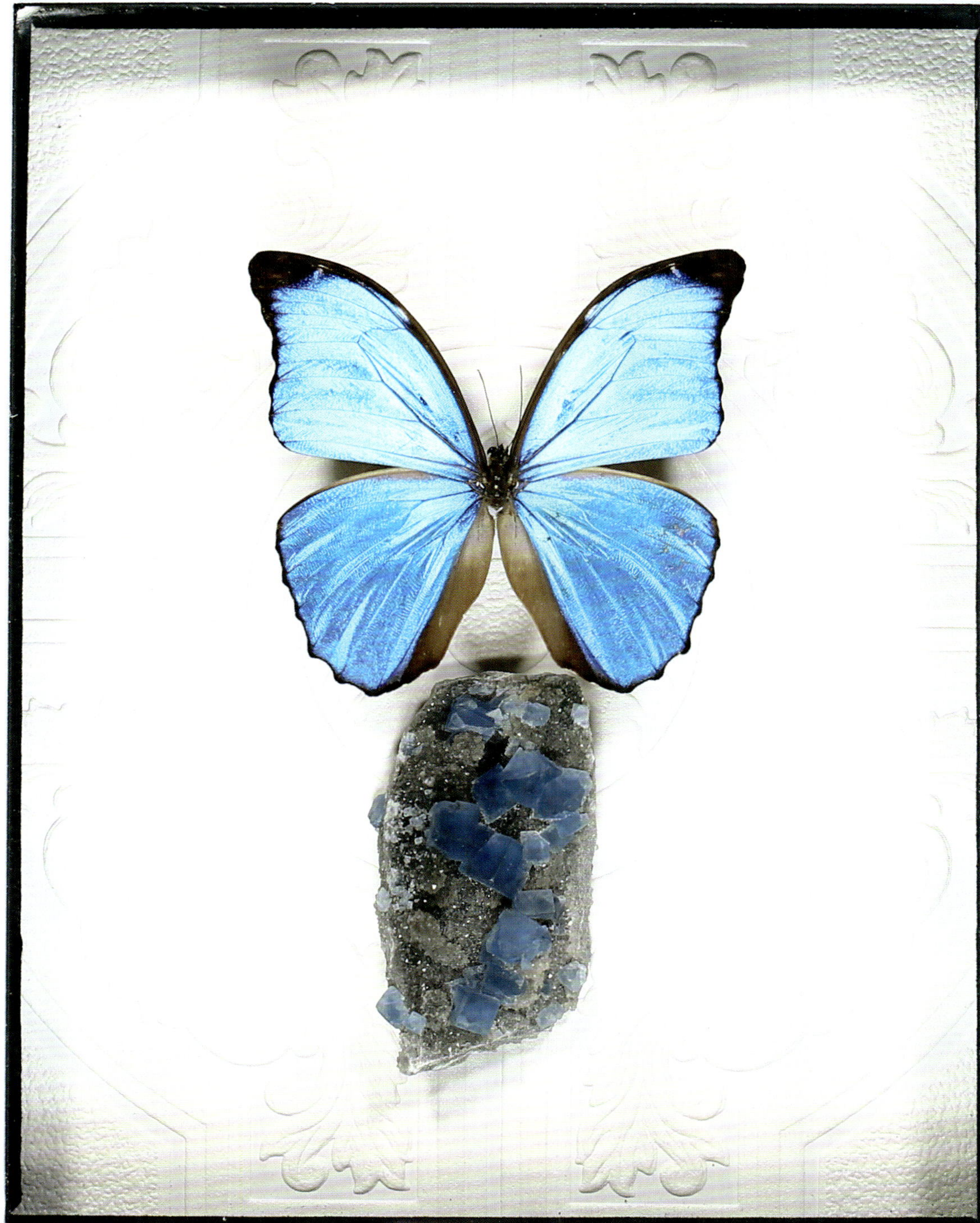

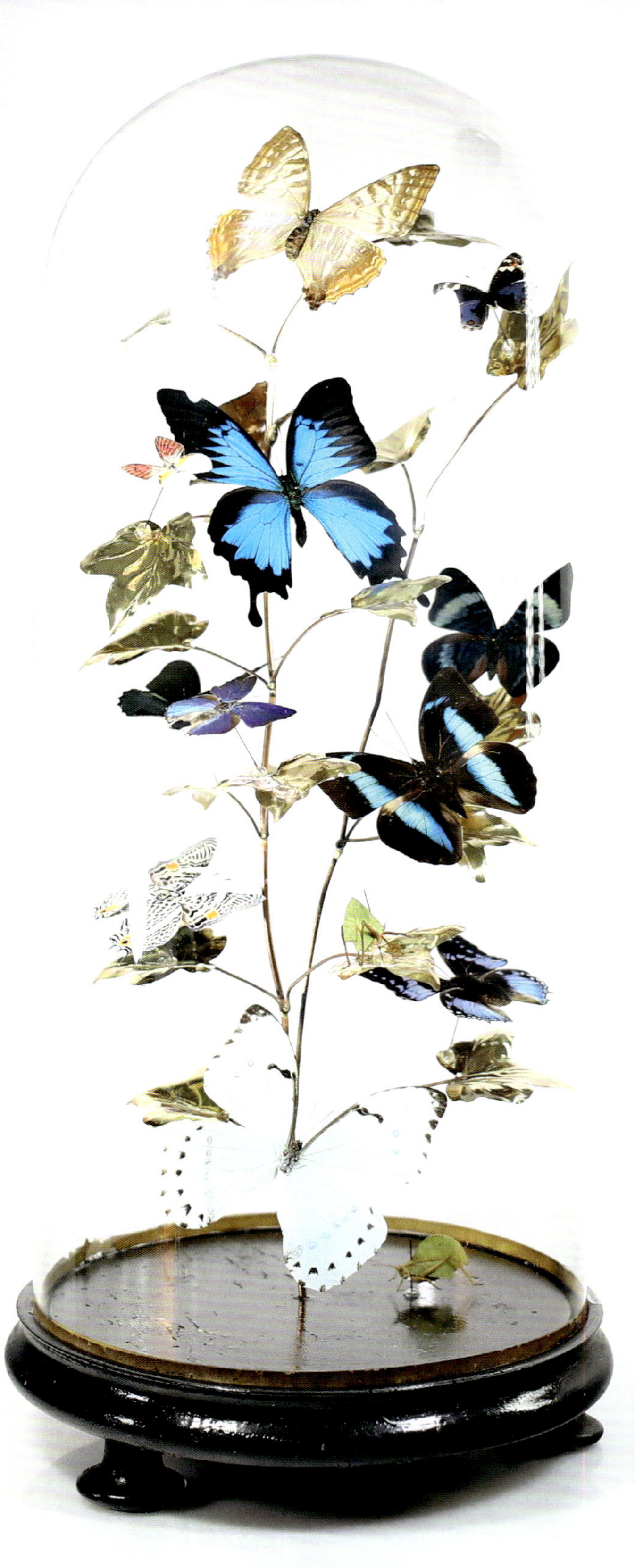

THE
GHOSTS
of NATURE

Translucence in Insects

In the intricate world of insects, the line that separates
visibility from invisibility is often beautifully blurred by the ethereal quality of
translucence. To the human eye, countless insects appear as fleeting shadows,
their forms partially obscured by a gauzy and luminous sheen that simultaneously
reveals and conceals their true shapes. This delicate veil is an evolved trait that
serves multiple purposes, including camouflage and communication. The transpar-
ent wings of a dragonfly, the gossamer exoskeleton of a cicada, or the glassy wings
of a moth—each displays a subtle elegance, a barely there armor that betrays
no obvious edges and no clear boundaries. From the faintest shimmer of a beetle's
exoskeleton to the almost imperceptible flight of a lacewing, translucence is an
evolutionary marvel that offers these remarkable creatures a significant advantage
in a world where survival often depends on remaining unseen—or on catching
the light at just the right moment.

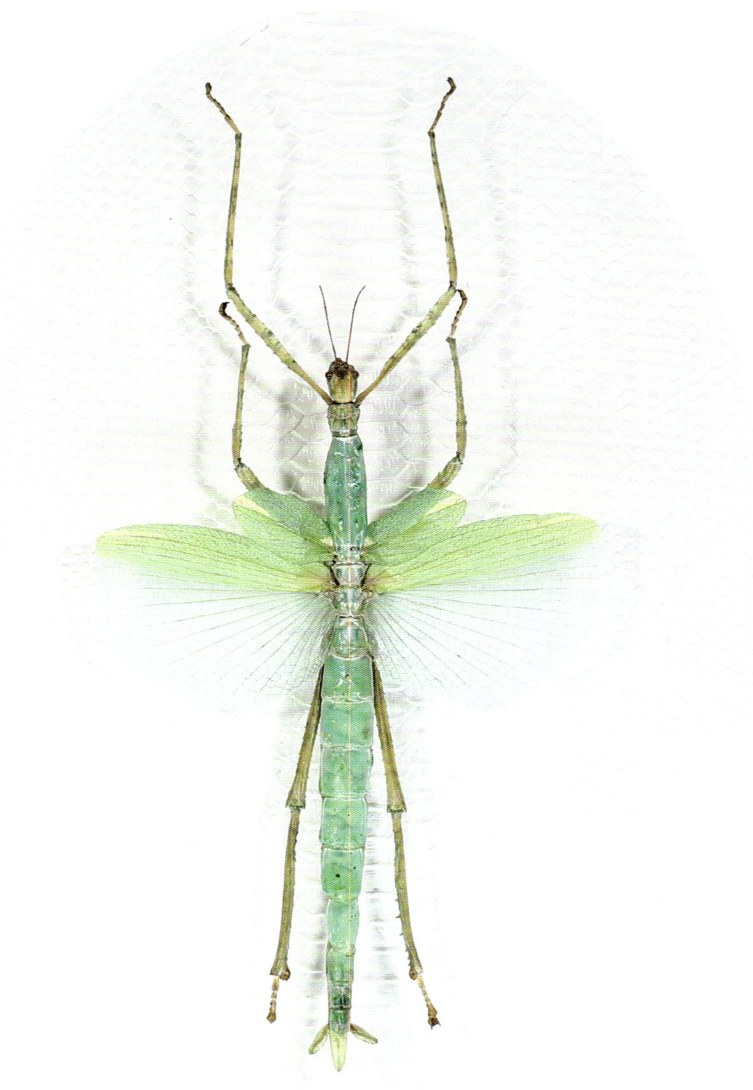

RED-WINGED GREEN GIANT STICK INSECT

SCIENTIFIC NAME
Eurycnema versirubra

GEOGRAPHIC RANGE
East Timor

This species of walking stick is often feared by those who encounter it in nature because of its size (females can be up to 12 inches [30.5 centimeters]), but it is actually harmless to humans and poses no real threat when encountered. If disturbed, females in this species use their hind legs to create a rustling noise that mimics the sound of a snake slithering through dry leaves.

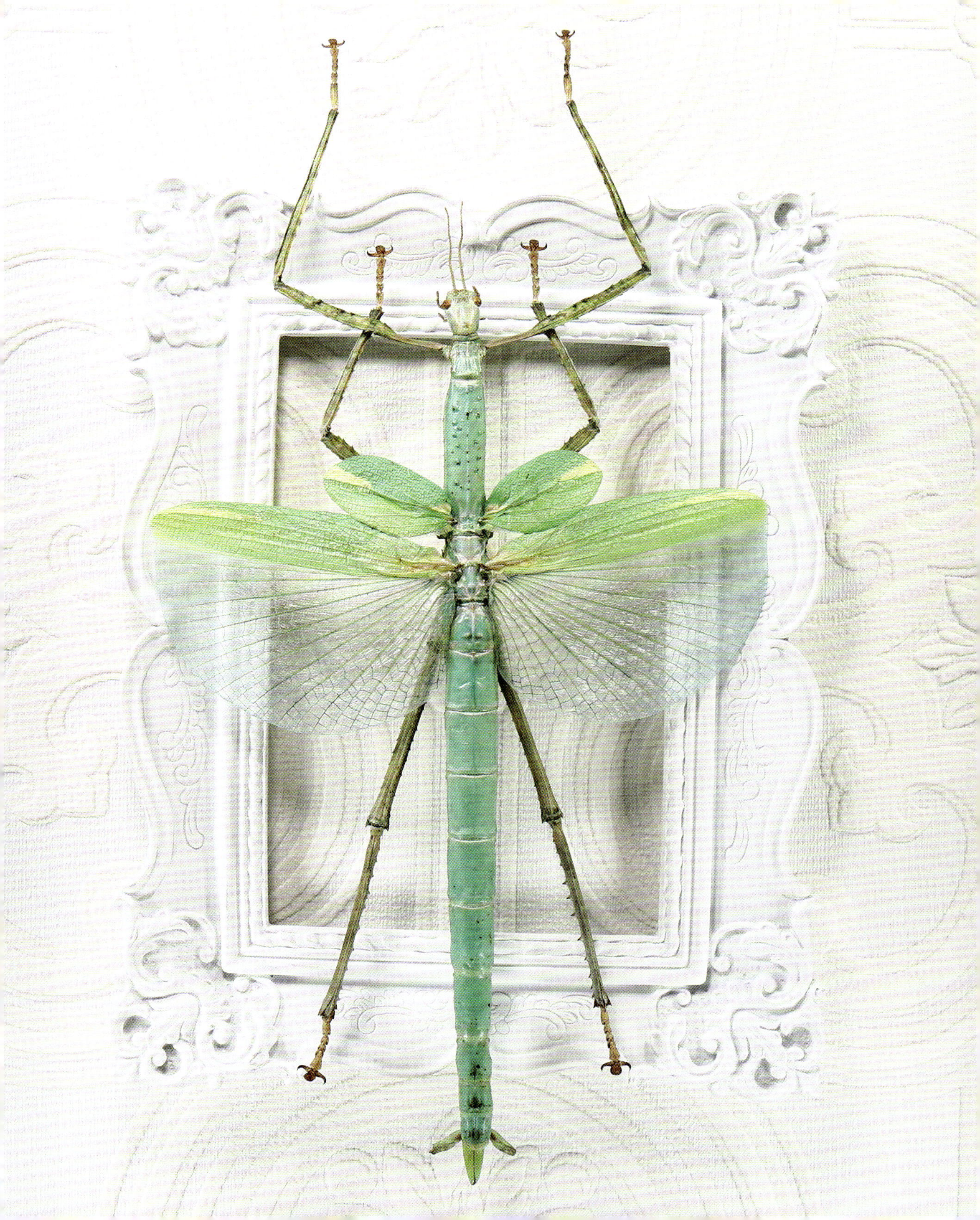

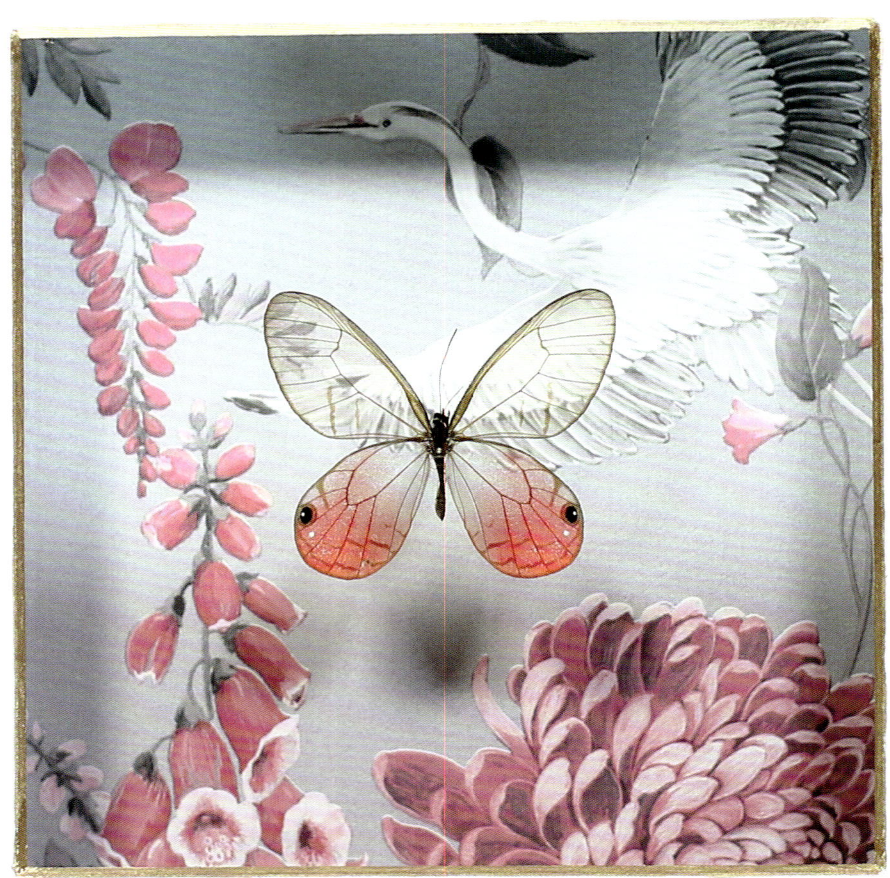

BLUSHING PHANTOM BUTTERFLY

SCIENTIFIC NAME
Cithaerias pireta

GEOGRAPHIC RANGE
Mexico, Central America,
and South America

These butterflies, known for their delicate appearance, have virtually no wing scales. As a result, observers can see the translucent wing membrane, which leads many collectors to call them glass-wing butterflies. Blushing phantoms also have a distinctive eyespot on the underside of each hind wing that some researchers believe resembles the head of a small snake, helping to trick and deter predators.

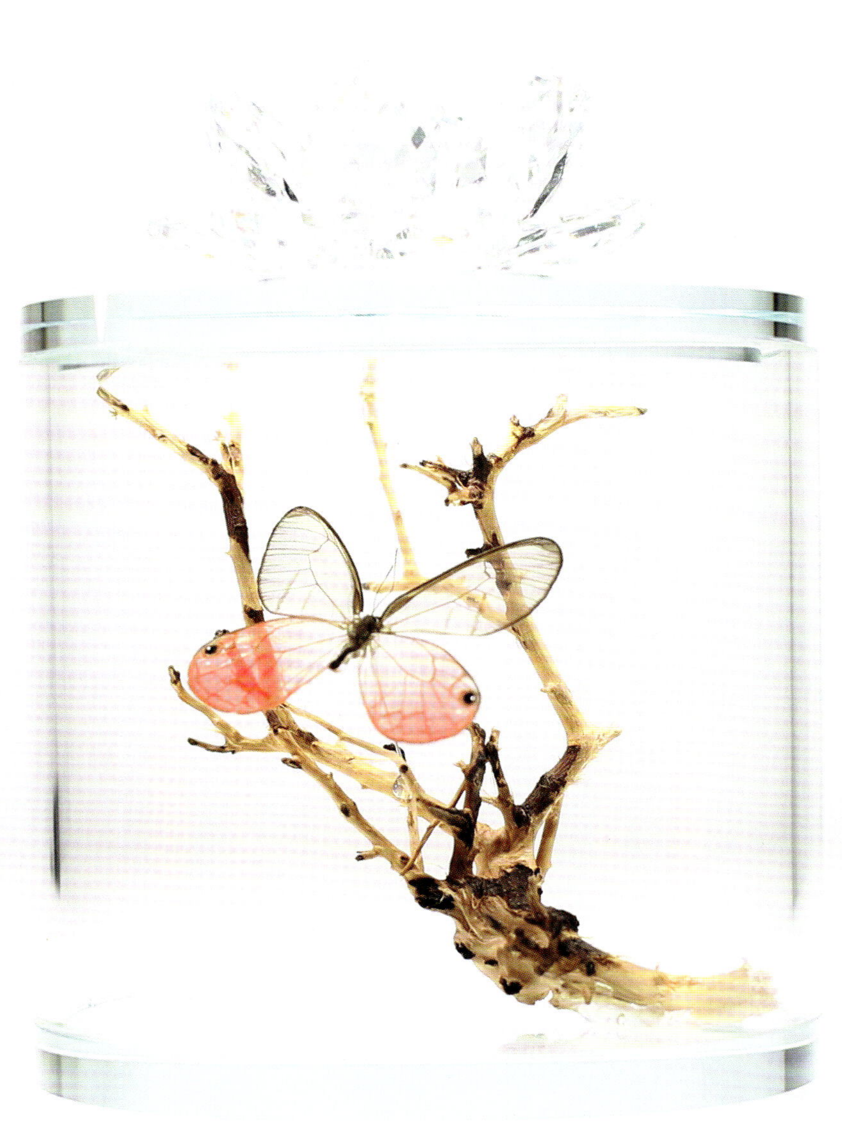

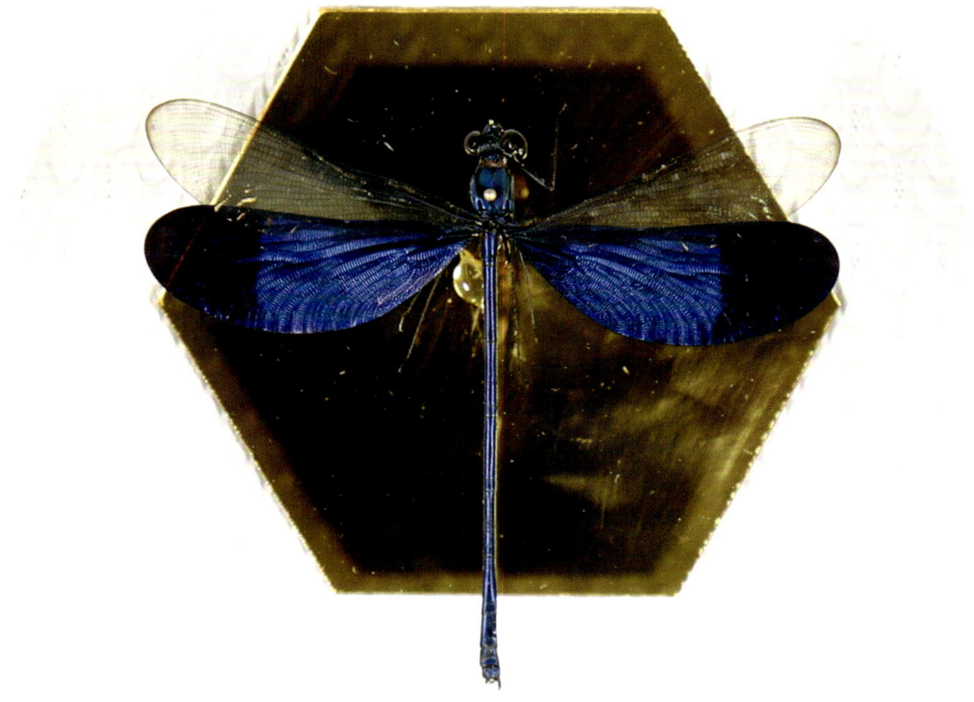

GREAT BLUE METALWING

SCIENTIFIC NAME
Neurobasis kaupi

GEOGRAPHIC RANGE
Indonesia

Native to the lush tropical rivers and streams of Indonesia, this species glistens in the sunlight. The males feature metallic blue hues, and the females exhibit brownish-yellow and green coloring. When a male intruder approaches too closely, protective males will take flight, soaring high into the air, where they flash their stunning iridescent wings to fend off the interloper.

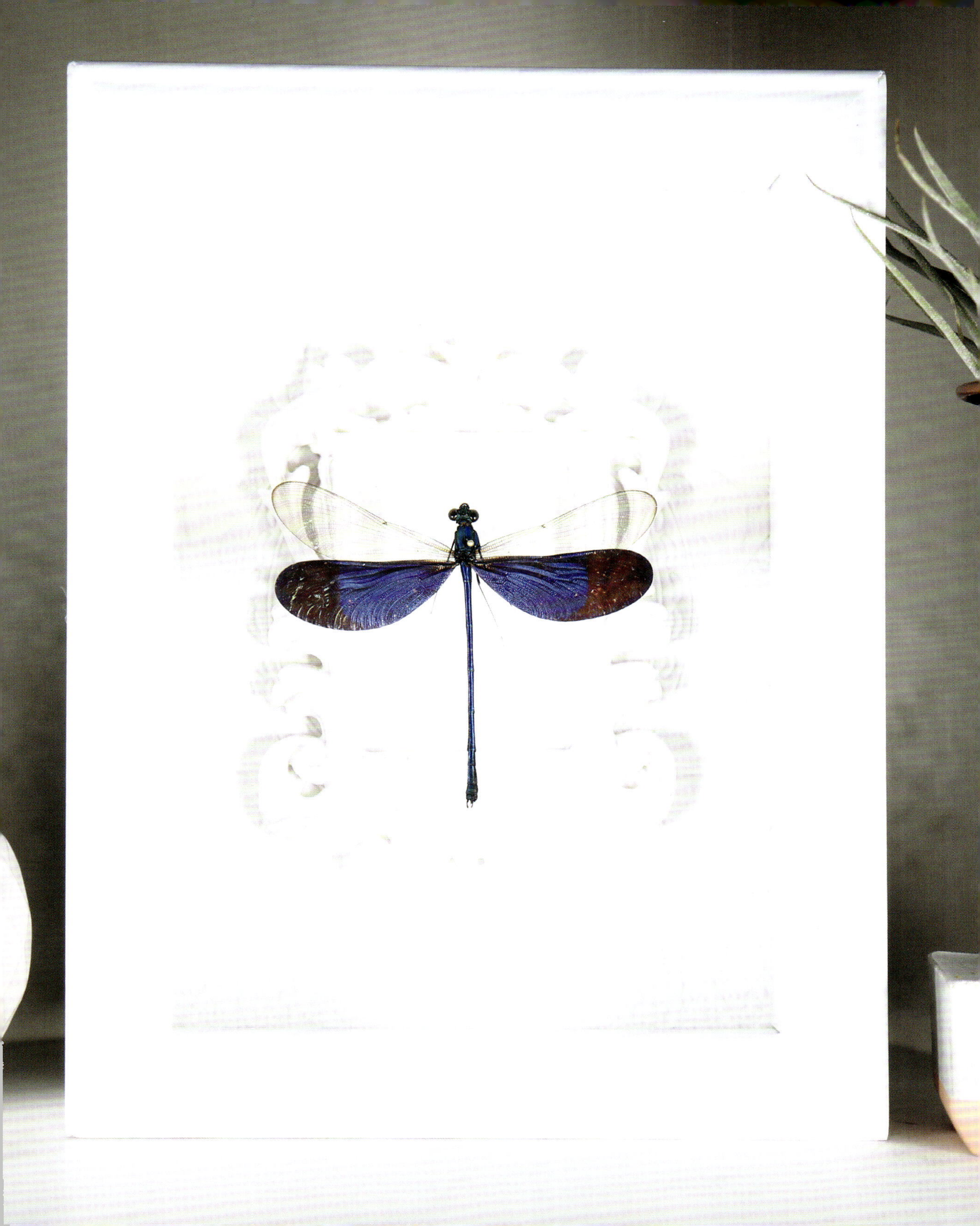

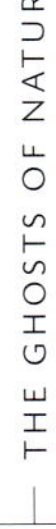

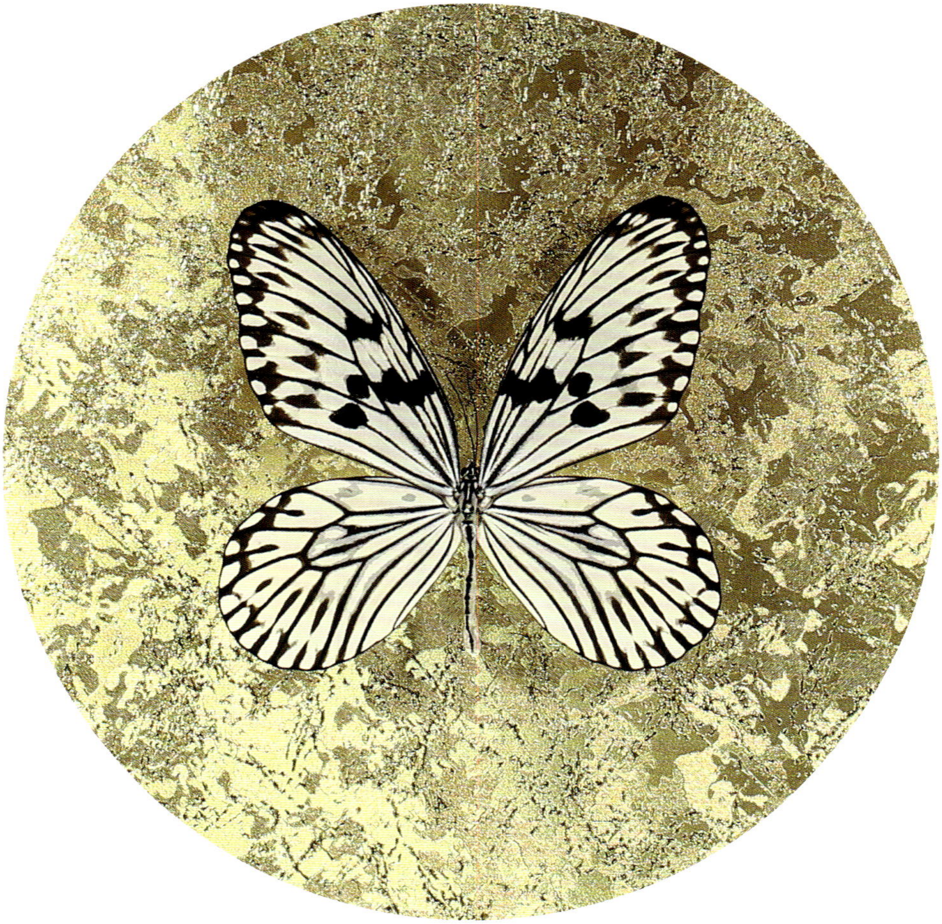

70

RICE PAPER BUTTERFLY

SCIENTIFIC NAME
Idea leuconoe

GEOGRAPHIC RANGE
Indonesia

This particular species is often present in butterfly houses and live butterfly expositions. Rice paper butterflies are known for their slow, exquisitely graceful flight, which doubles as an effective defense mechanism: Birds have developed an aversion to gliding butterflies like these, as they are toxic, whereas quick-moving ones are usually not.

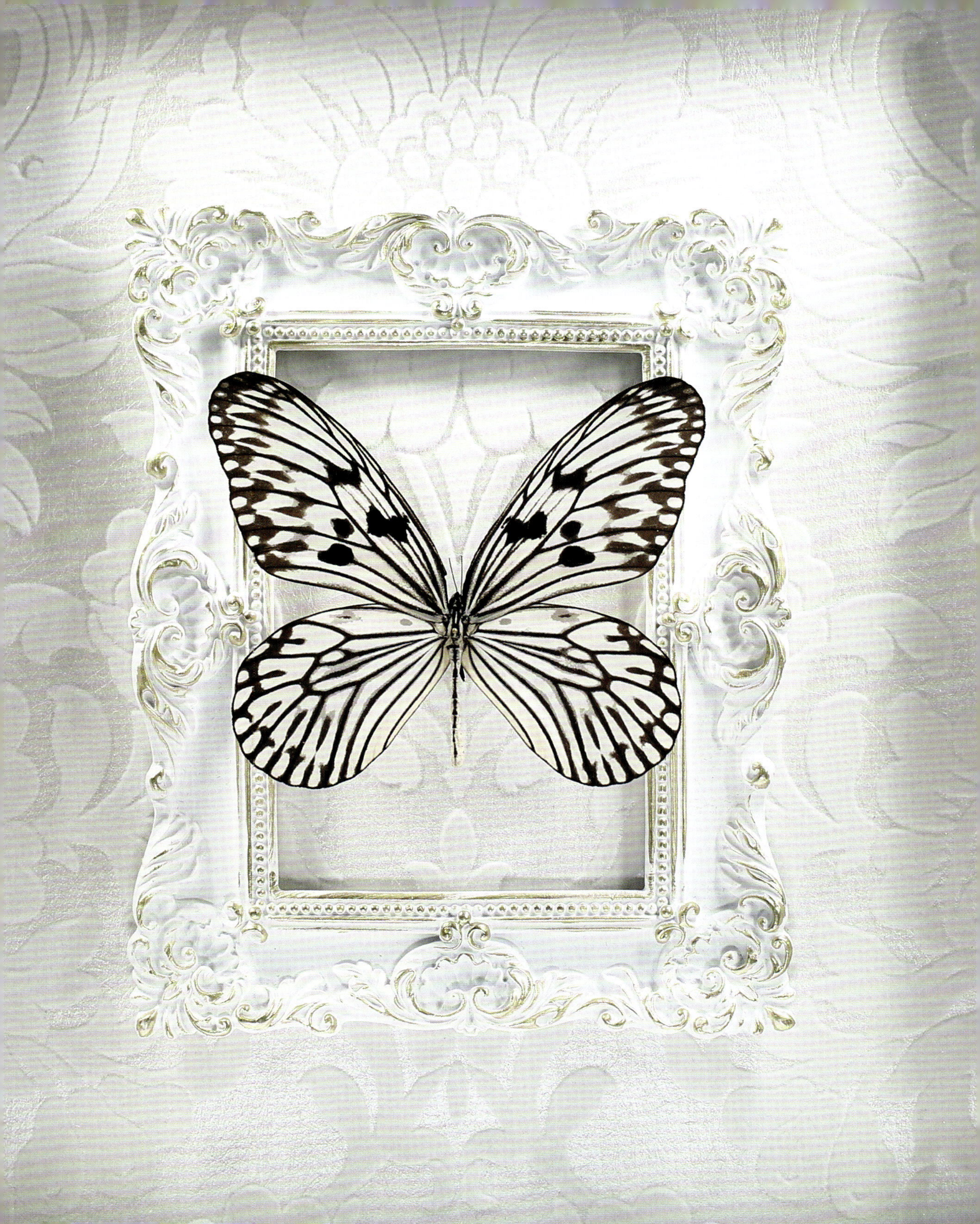

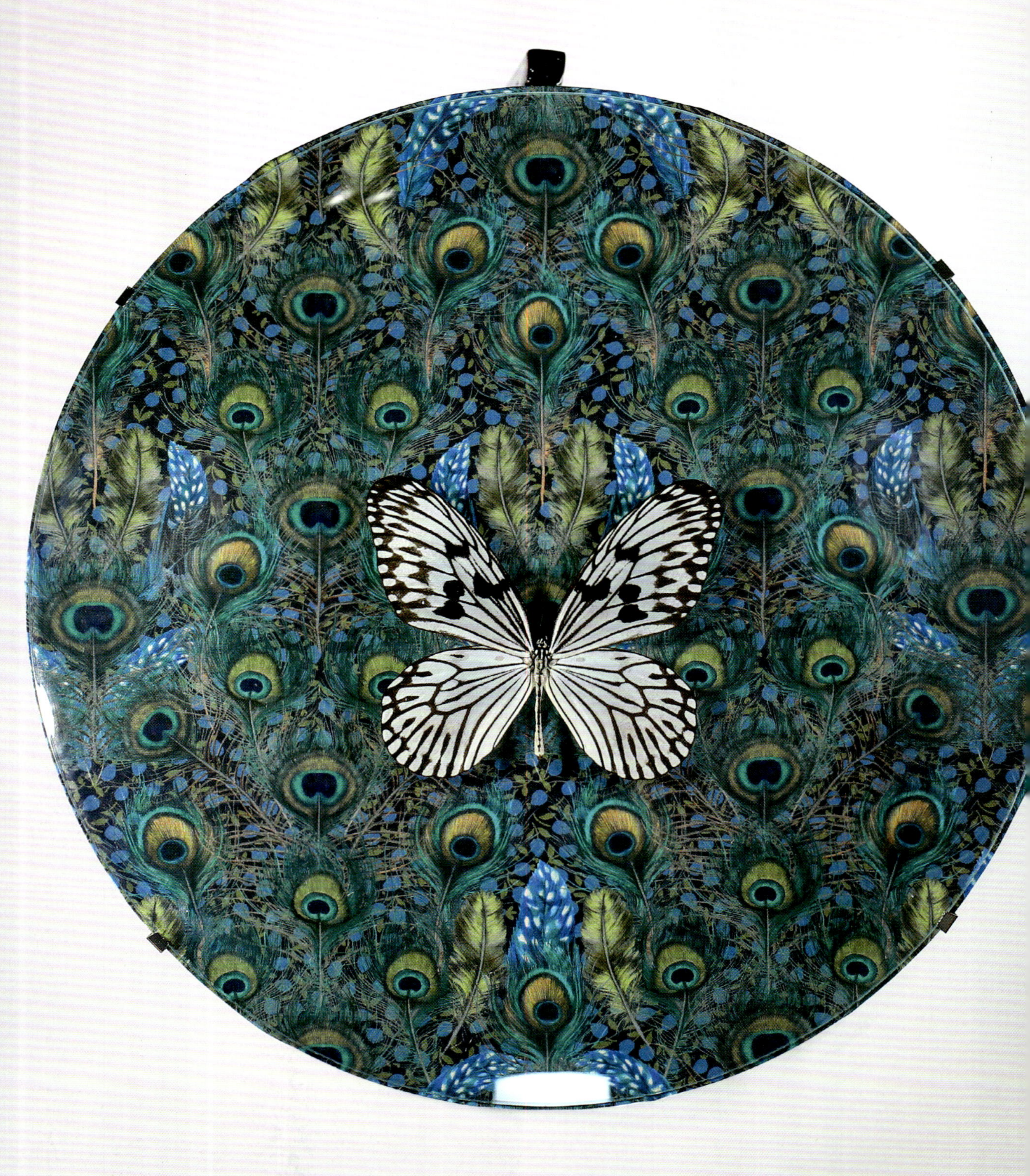

GREEN DARNER DRAGONFLY

SCIENTIFIC NAME
Anax junius

GEOGRAPHIC RANGE
North America

Darners are skilled at catching and consuming various insects, including common pests such as mosquitoes and flies. Even in their aquatic larval stage, Darner nymphs catch and eat other aquatic insects using their labial mask—a sharp device that extends from below their mandibles. Their role in controlling insect populations makes them an important part of their local ecosystems.

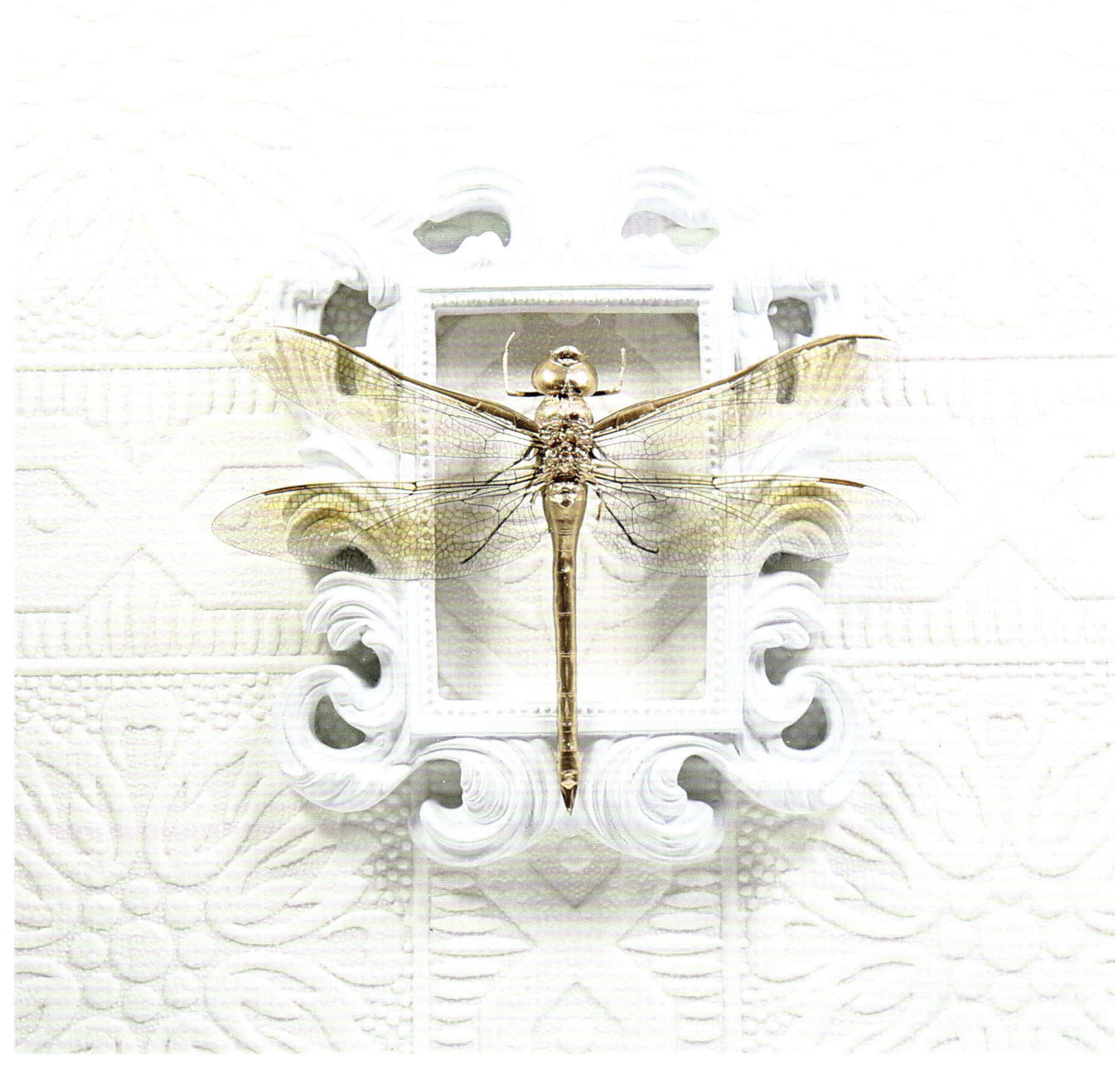

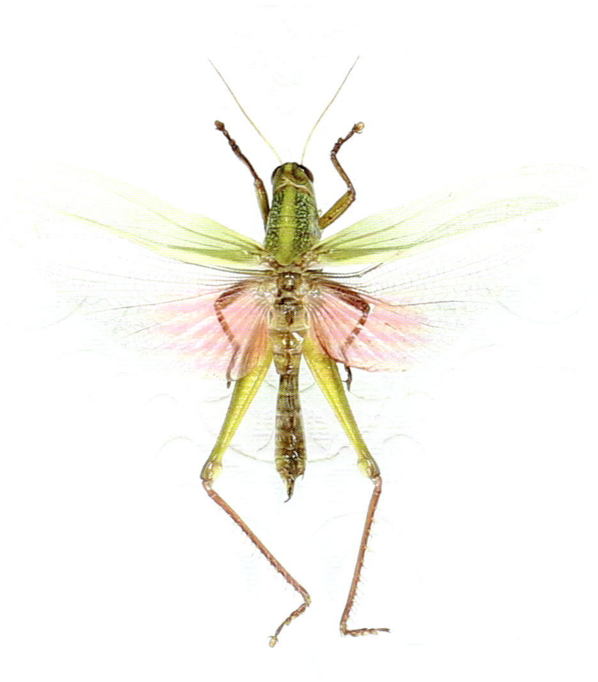

CITRUS LOCUST

SCIENTIFIC NAME
Chondracris rosea

GEOGRAPHIC RANGE
India, East Asia, and
Southeast Asia

This particular species of locust is recognized as the largest variety found in China. Like other large locusts, these creatures are surprisingly agile and can leap impressive distances, jumping up to twenty times their own body length with ease.

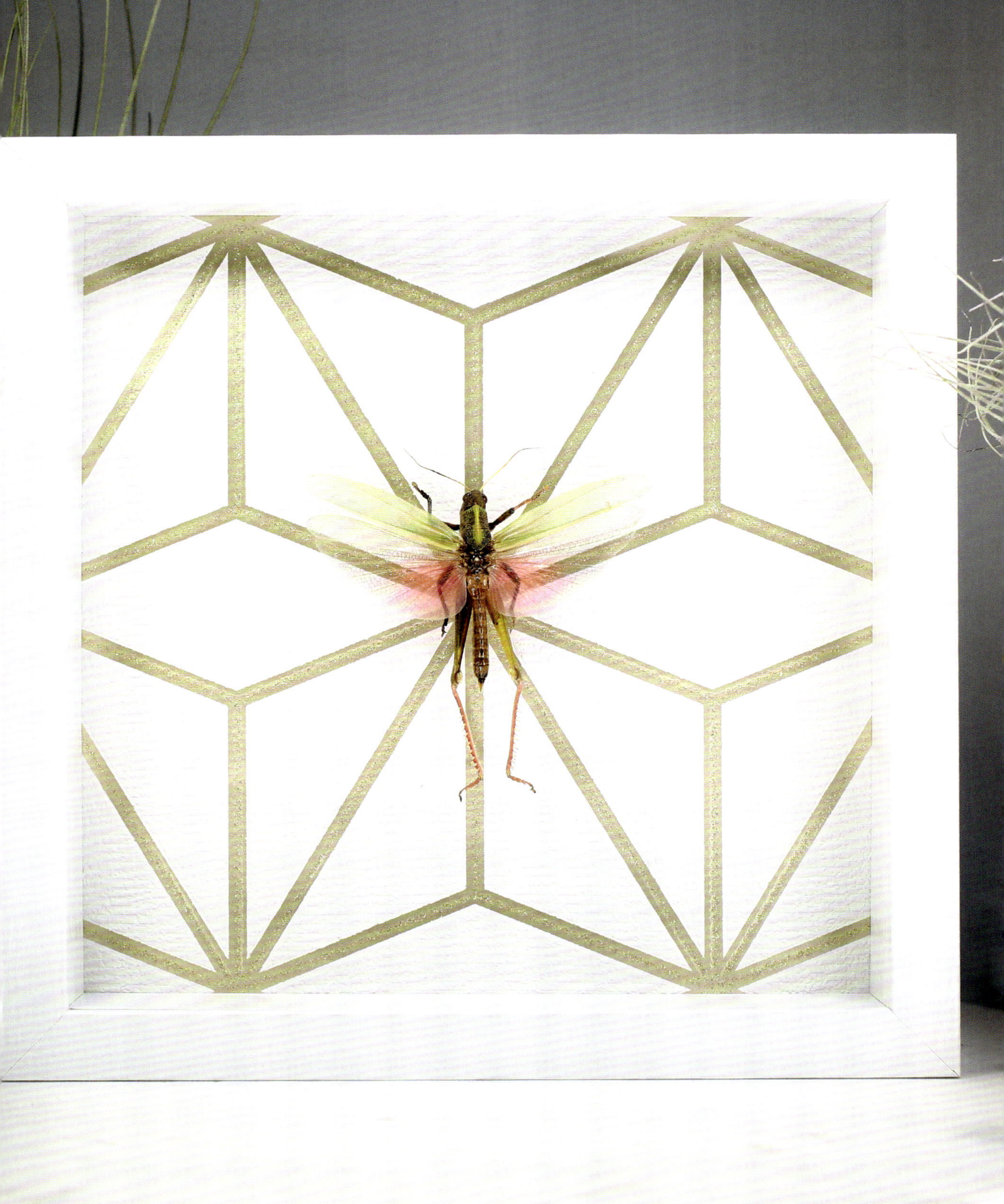

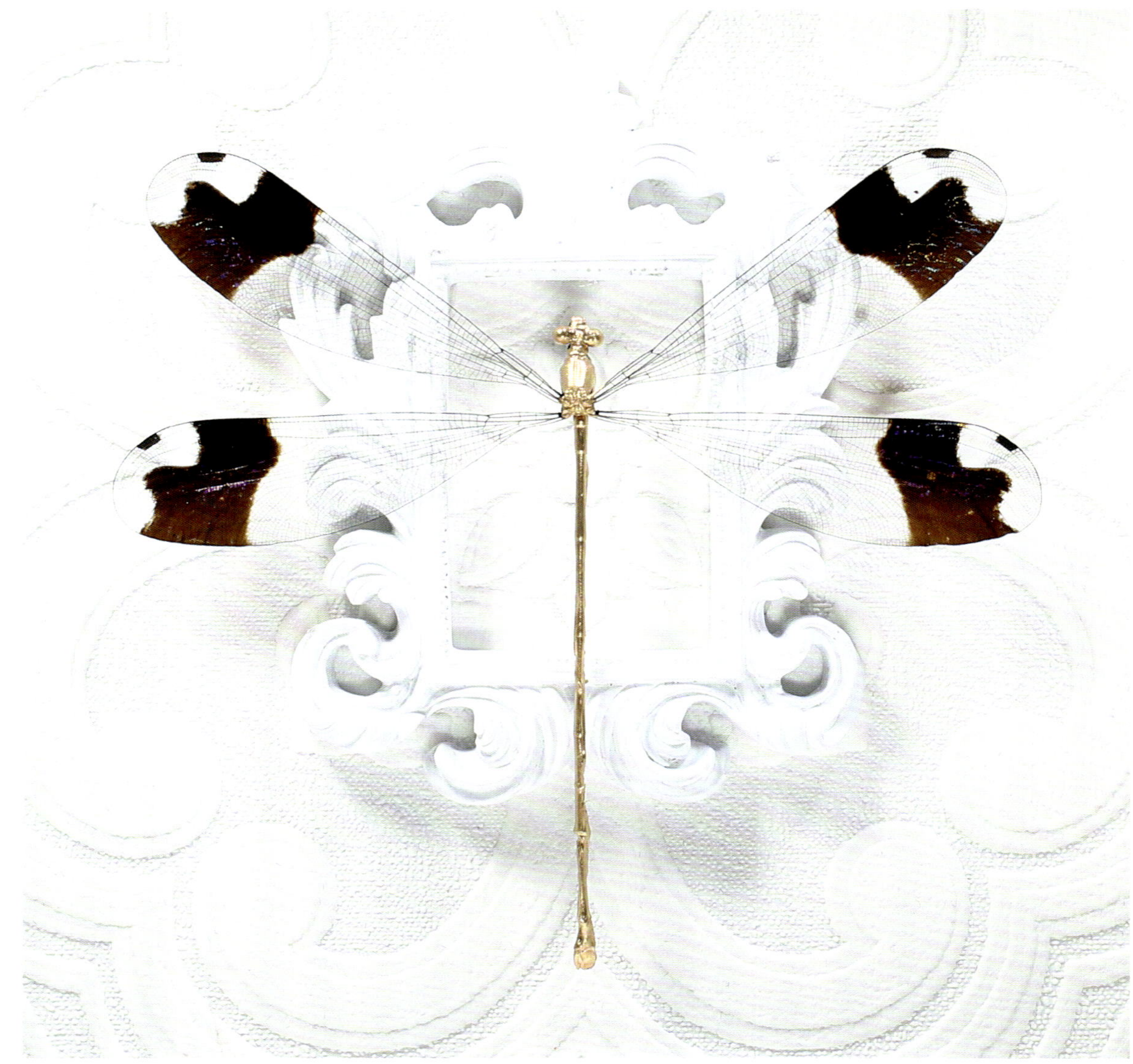

BLUE-WINGED HELICOPTER DAMSELFLY

SCIENTIFIC NAME
Megaloprepus caerulatus

GEOGRAPHIC RANGE
Mexico, Central America,
and South America

These damselflies rank among the largest species globally, with a wingspan of up to 7.5 inches (19 centimeters). They primarily prey on orb weavers dwelling in the forest understory, skillfully extracting them from their intricate webs. Unlike many damselfly and dragonfly species, in which females are typically larger, male helicopter damselflies are typically the larger sex.

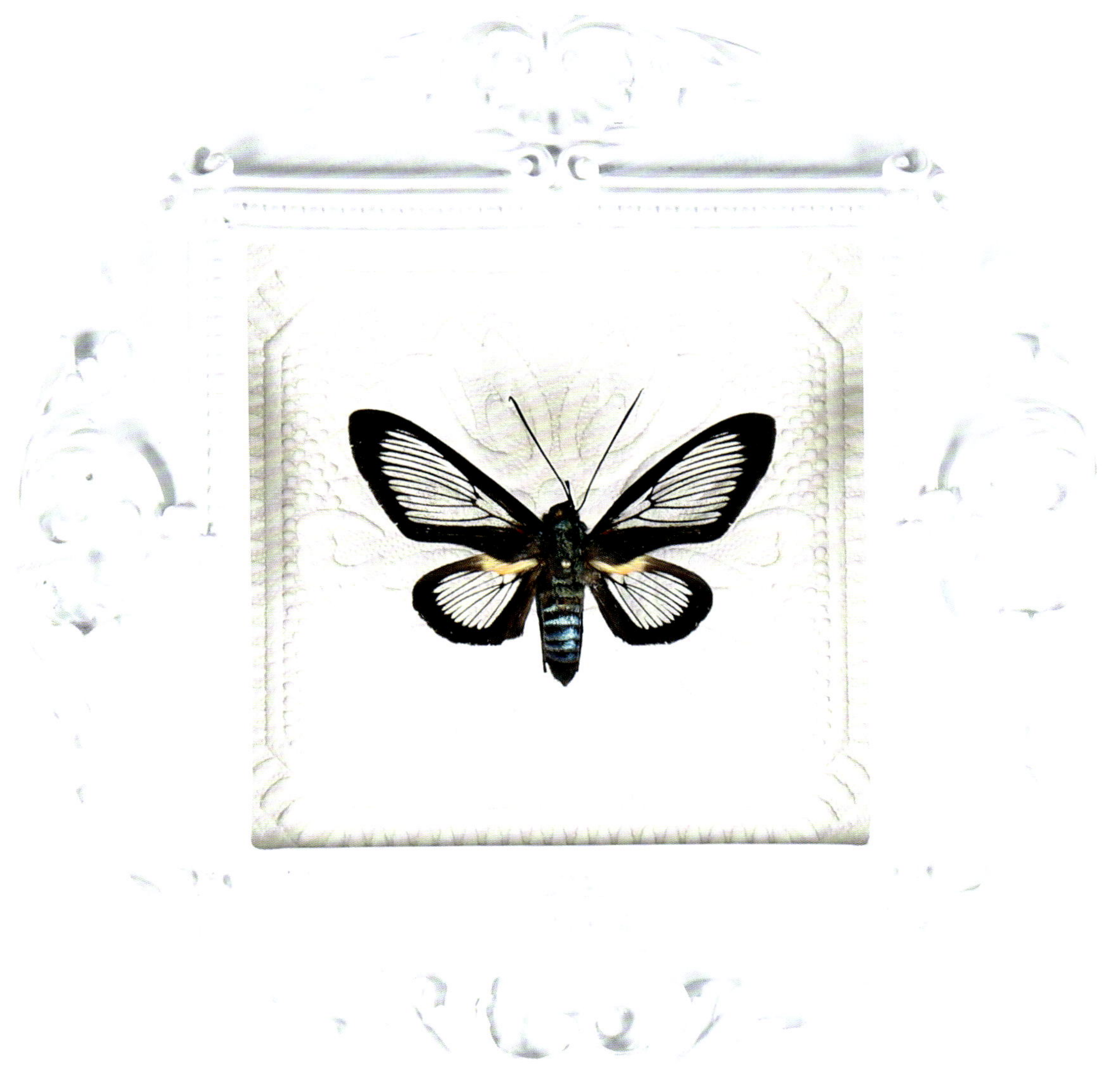

COCYTIA MOTH

SCIENTIFIC NAME
Cocytia durvillii

GEOGRAPHIC RANGE
Moluccas, Aru Islands,
and New Guinea

Cocytia moths are well known for being active during the day and showing behavioral convergence with butterflies—including flower feeding, sun basking, and rapid flight through the open forest.

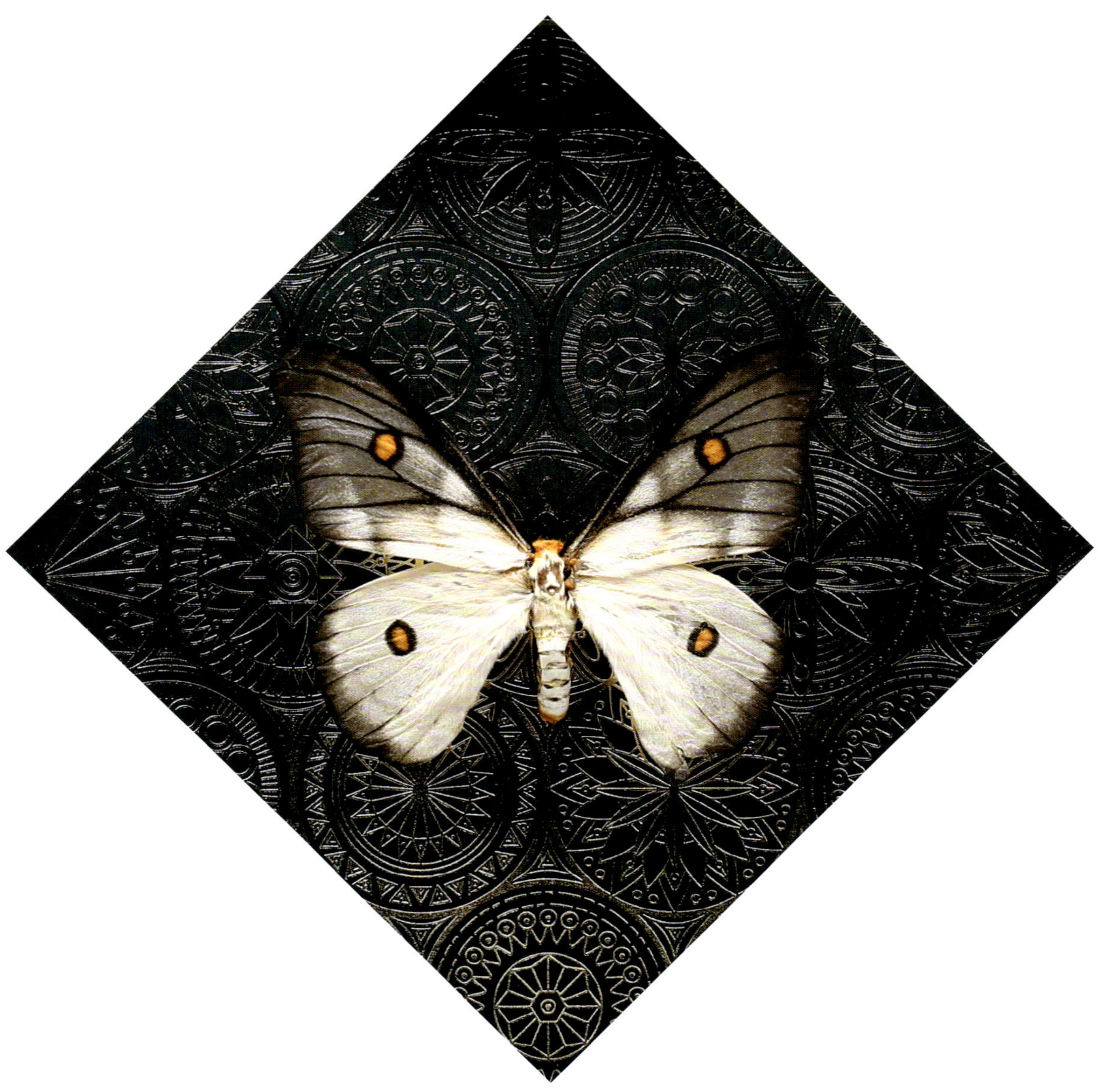

GHOSTLY SILKMOTH

SCIENTIFIC NAME
Ceranchia apollina

GEOGRAPHIC RANGE
Madagascar

As adults, members of this species survive solely on the lipids they stored during their larval stages. Because the ghostly silkmoth lacks a digestive tract and has a vestigial mouth, it can survive for only a few days as an adult. The beautiful gold silk produced by this moth is strong.

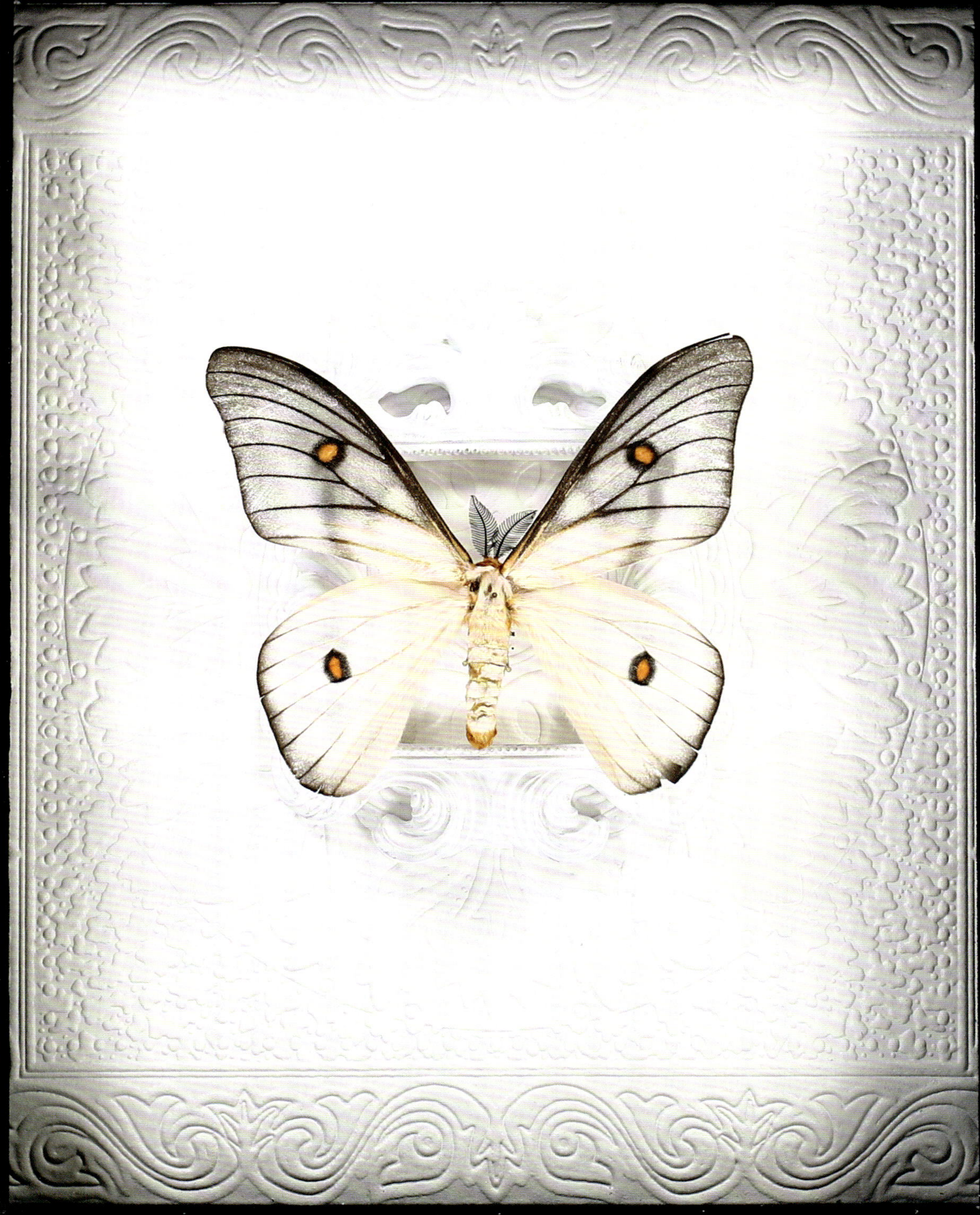

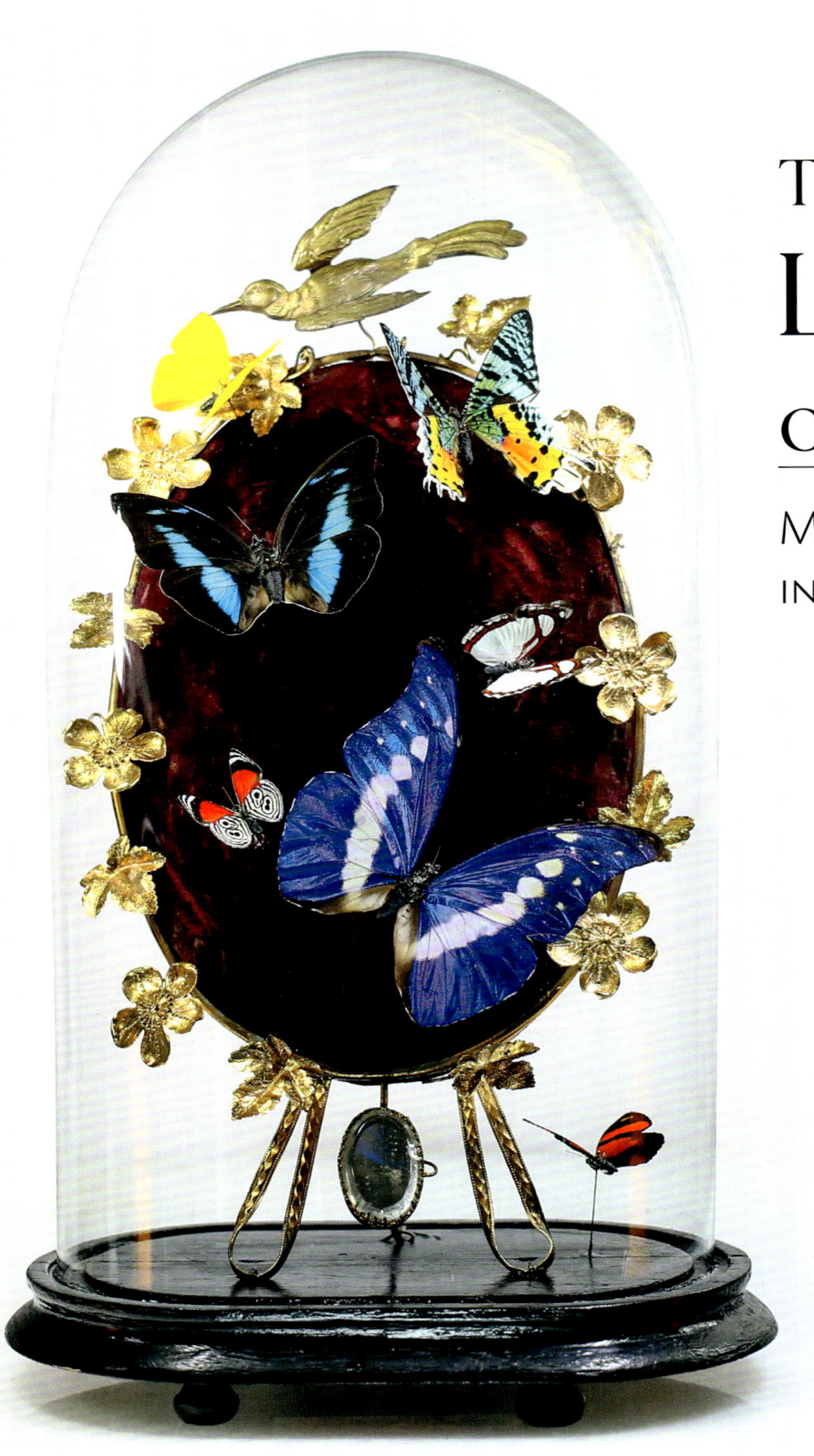

THE LUSTER of LIFE

Metallic Shine in Insects

IN THE DEPTHS OF THE VAST NATURAL WORLD, WHERE THE SMALLEST details often hold the greatest and most profound mysteries, one notable phenomenon stands out conspicuously in its sheer brilliance: the captivating metallic shine of certain intriguing insects. To witness an insect glowing with this otherworldly luster is akin to catching a fleeting glimpse of nature's alchemy—where light, color, and form seamlessly merge into an iridescent jewel of exquisite beauty. From the striking emerald armor of a jewel beetle to the radiant golden gleam of a fly's wings, the metallic shine of some insects helps them dazzle, deceive, and endure in the face of challenges. It acts as a formidable shield, warding off potential predators with an otherworldly glow that helps them appear larger and more dangerous than they really are. In some cases, it attracts potential mates with a brilliance that effectively signals health, genetic fitness, or territorial claims. Metallic luster is a testament to nature's intricate design, whereby creatures harness the power of light itself.

METALLIC STAG BEETLE

These beetles exhibit extreme sexual dimorphism, characterized by males being significantly larger than their female counterparts and possessing remarkably long mandibles. This curious trait has made these beetles the subject of various genetic studies aimed at determining the specific genes involved in the development and expression of pronounced sexual dimorphism.

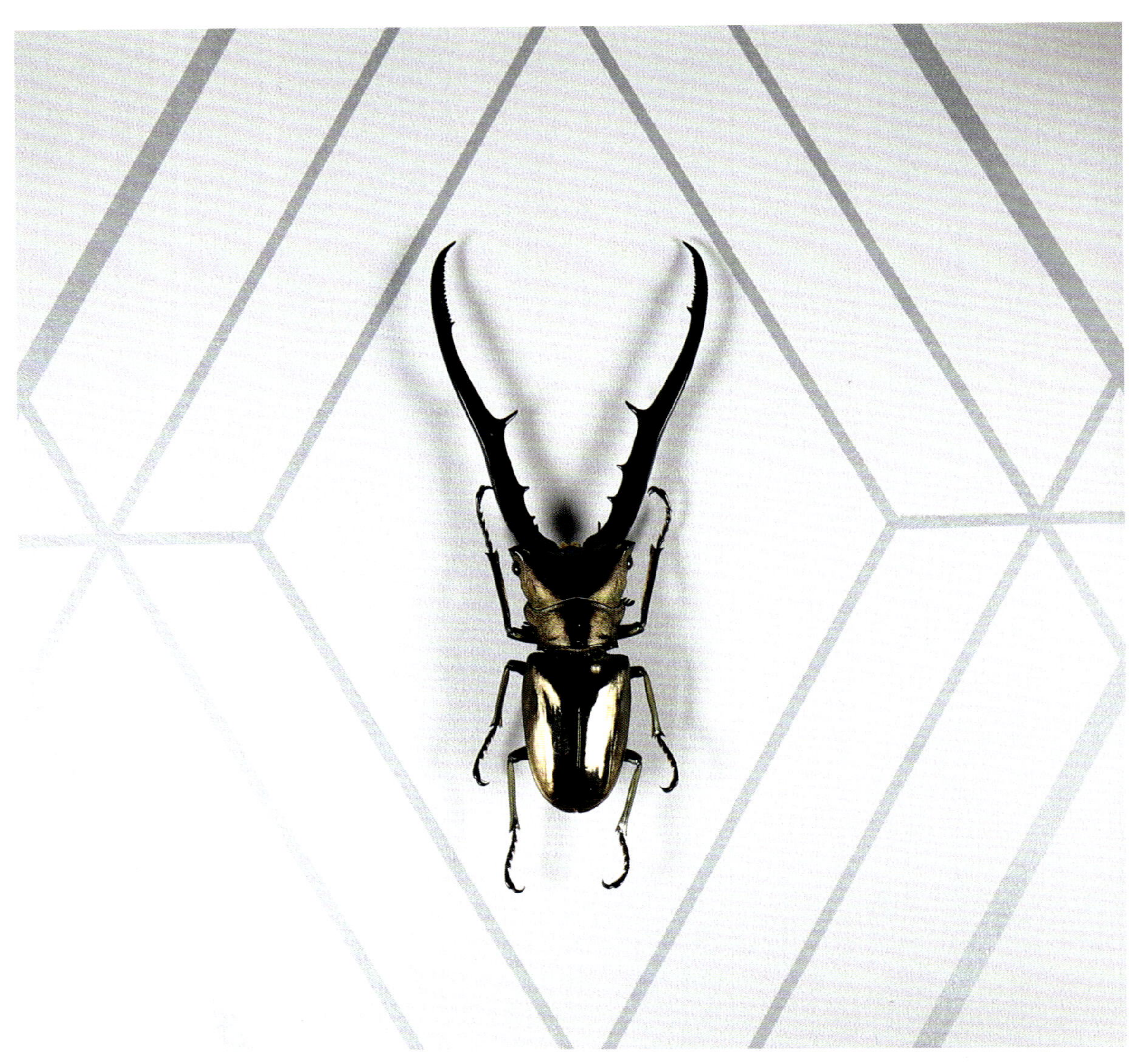

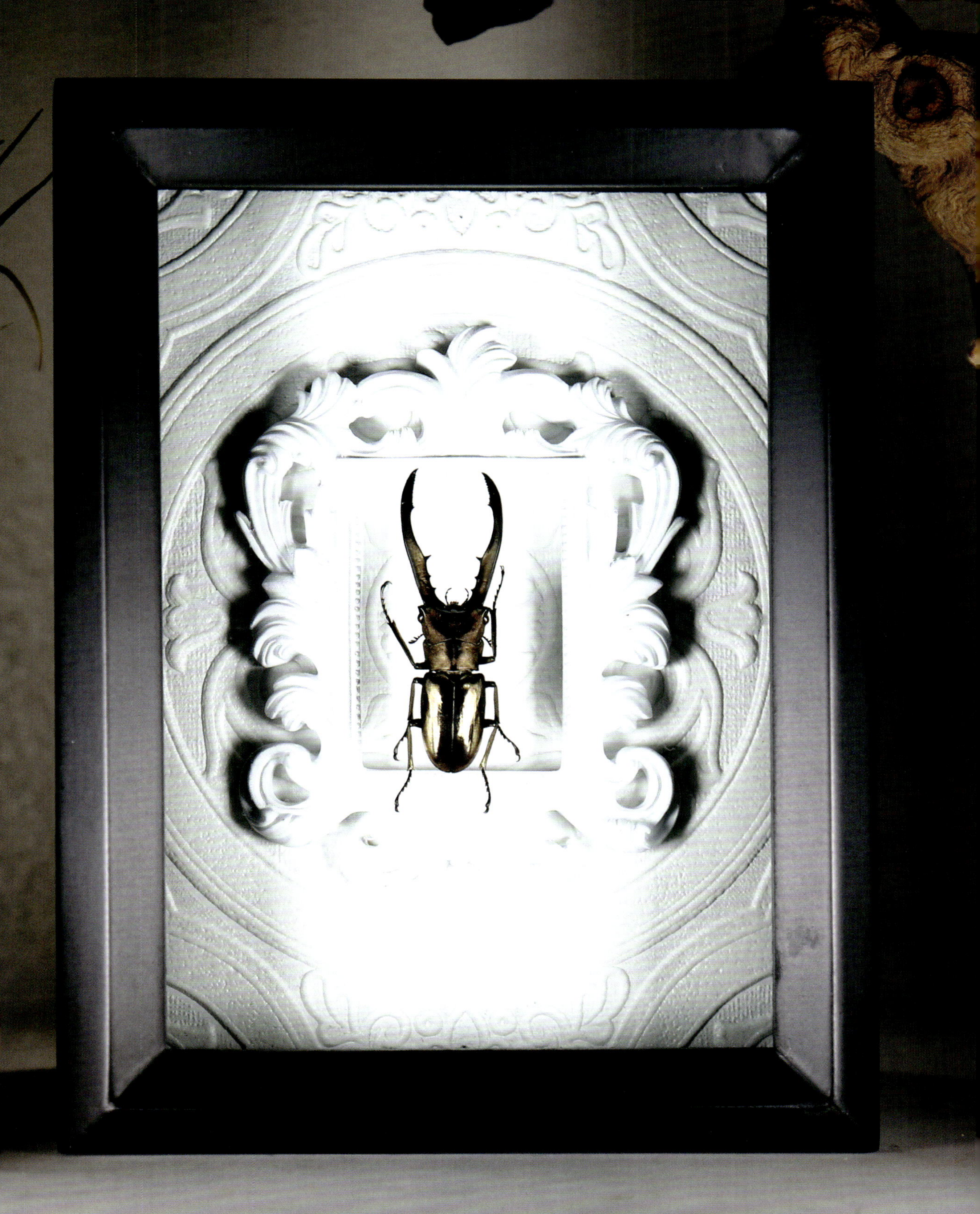

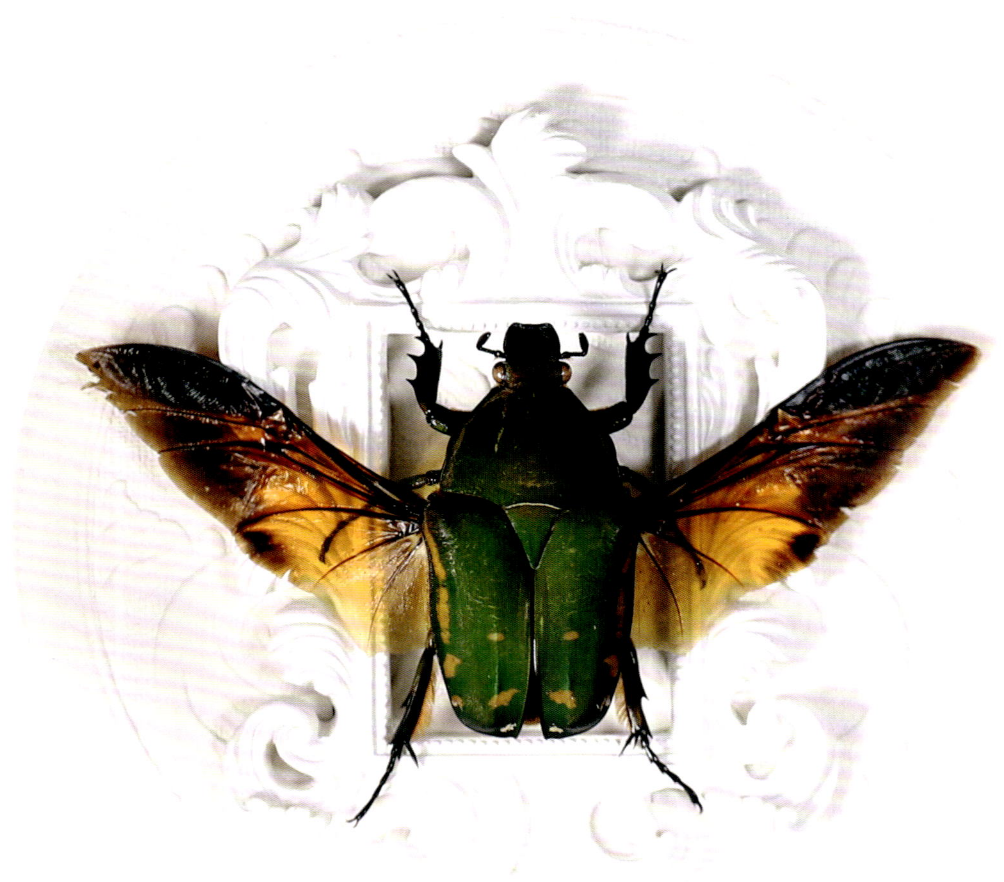

GIANT FLOWER BEETLE

SCIENTIFIC NAME
Mecynorhina torquata

GEOGRAPHIC RANGE
Cameroon, Democratic
Republic of the Congo,
and Uganda

This species is among the largest flower beetles in the world, second only to goliath beetles. The adult beetles of this African species feed primarily on nectar from various flowering plants and on fallen fruit. During their larval stages, however, they survive on composted leaves and decomposing wood, which play a crucial role in their development.

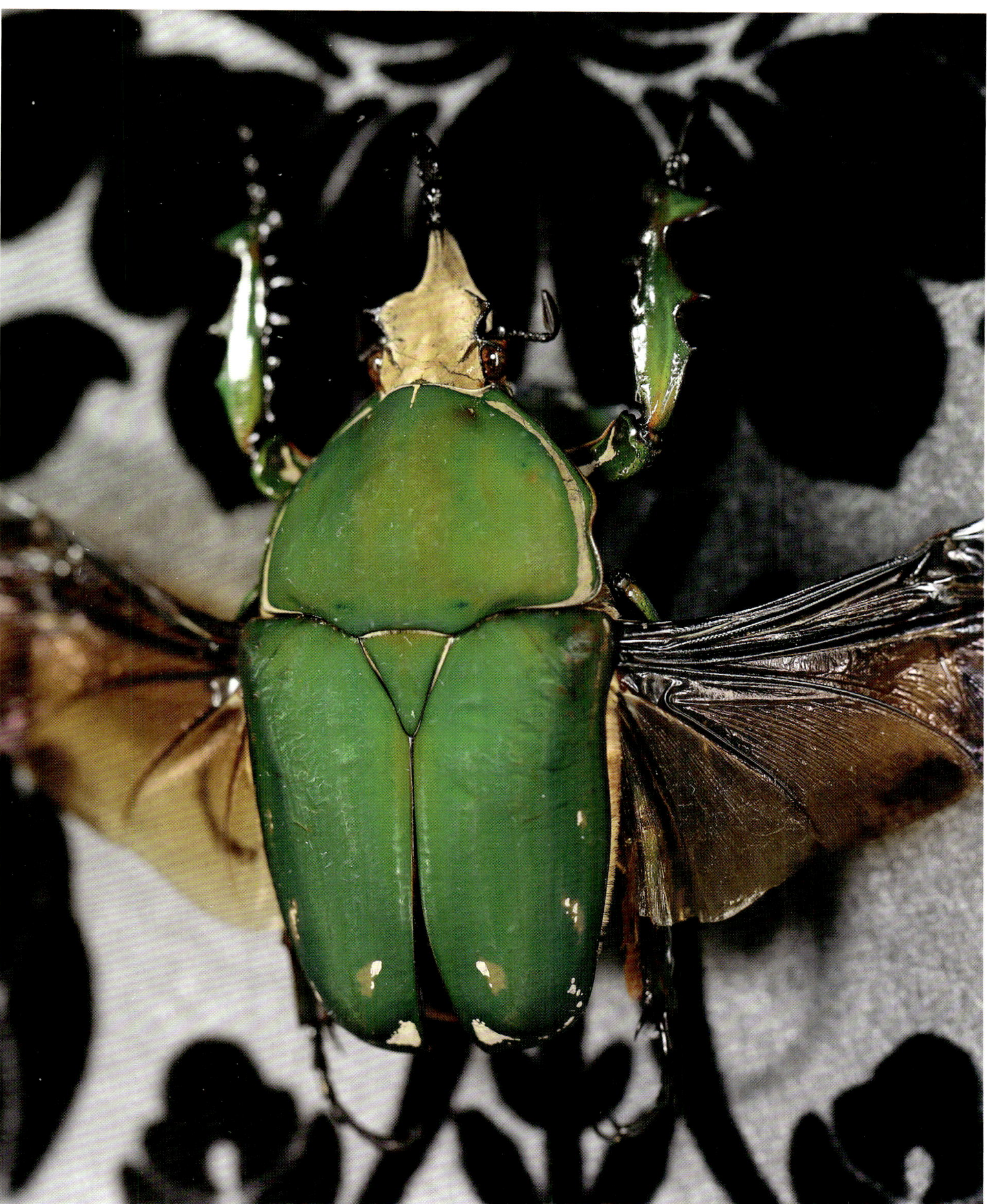

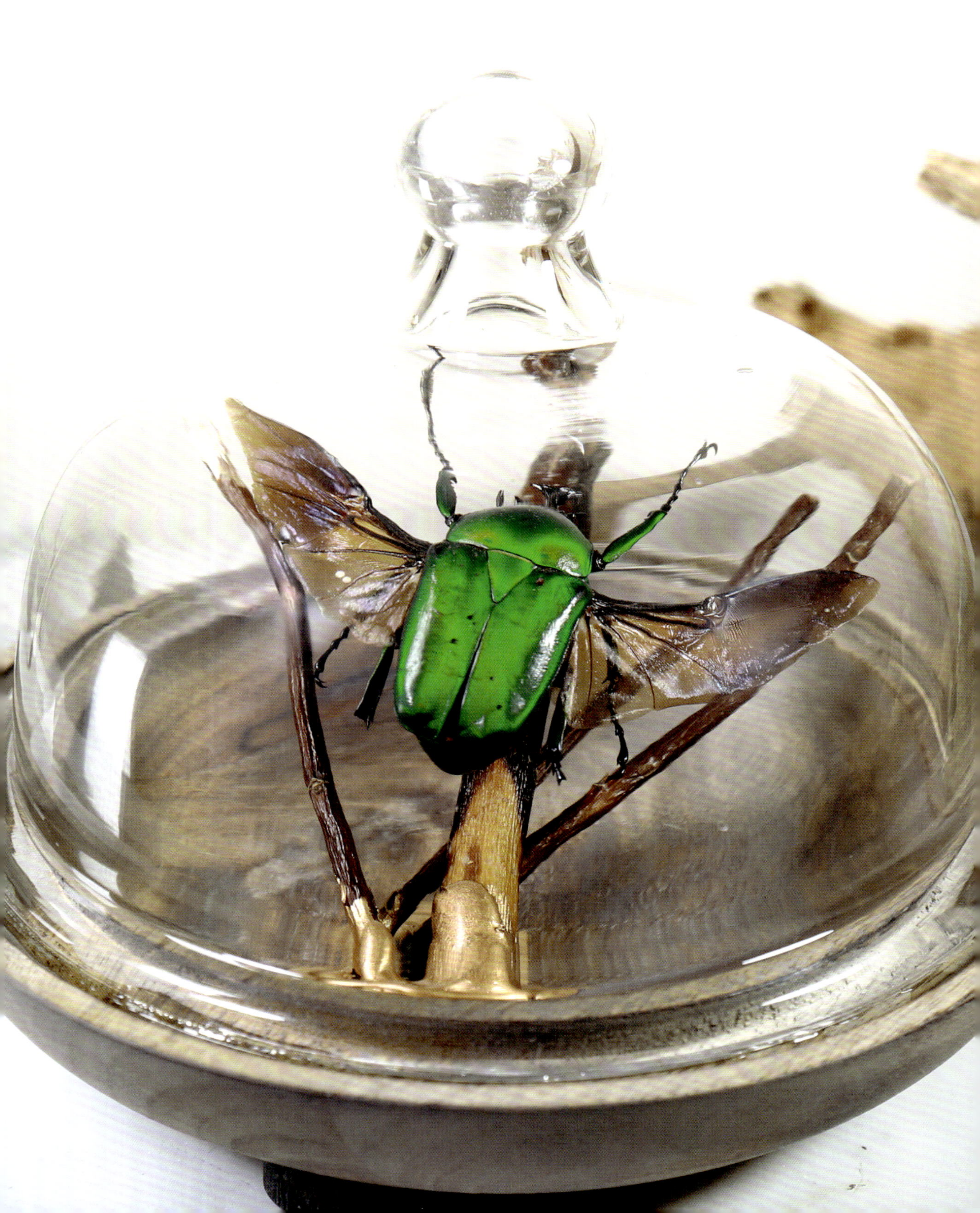

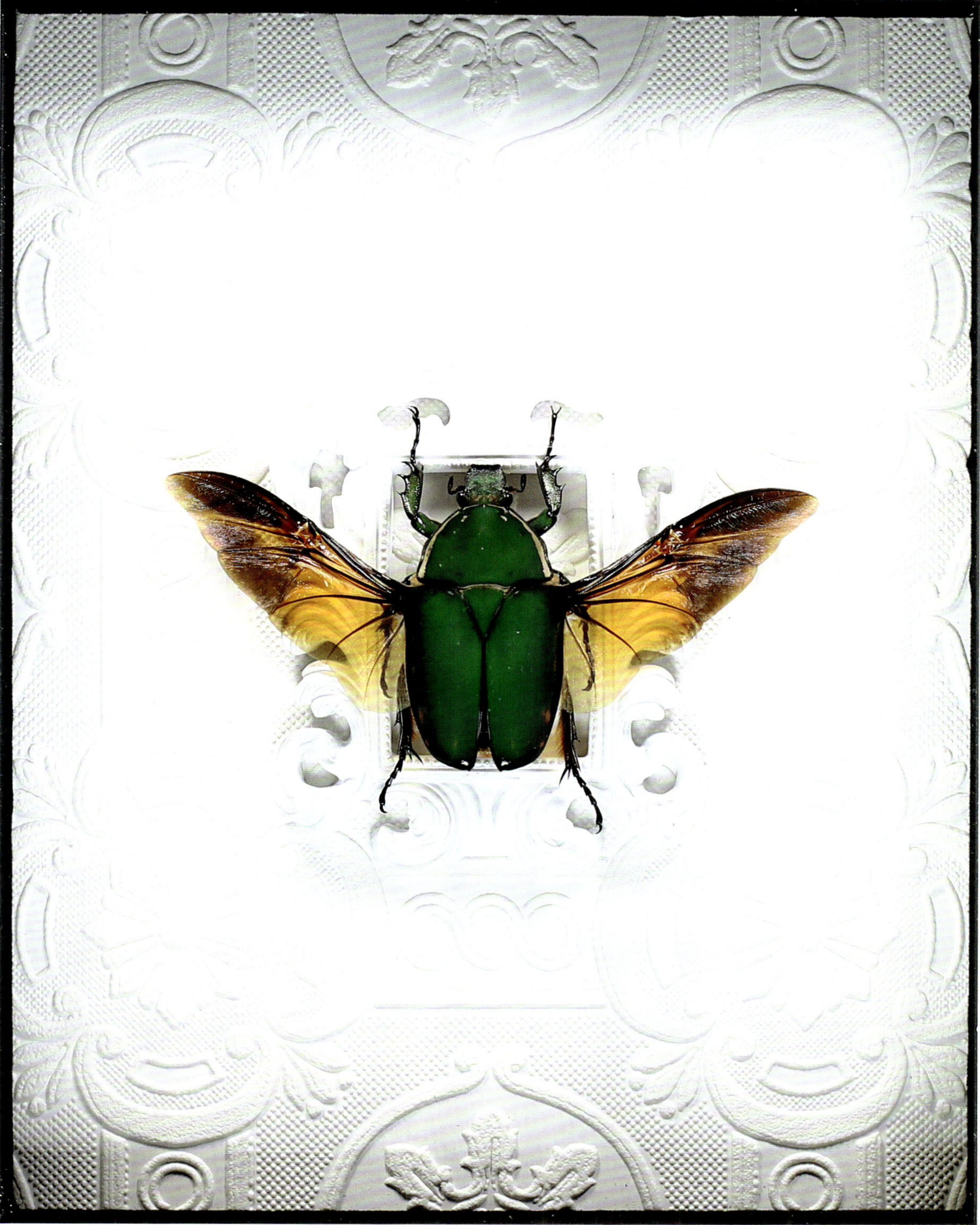

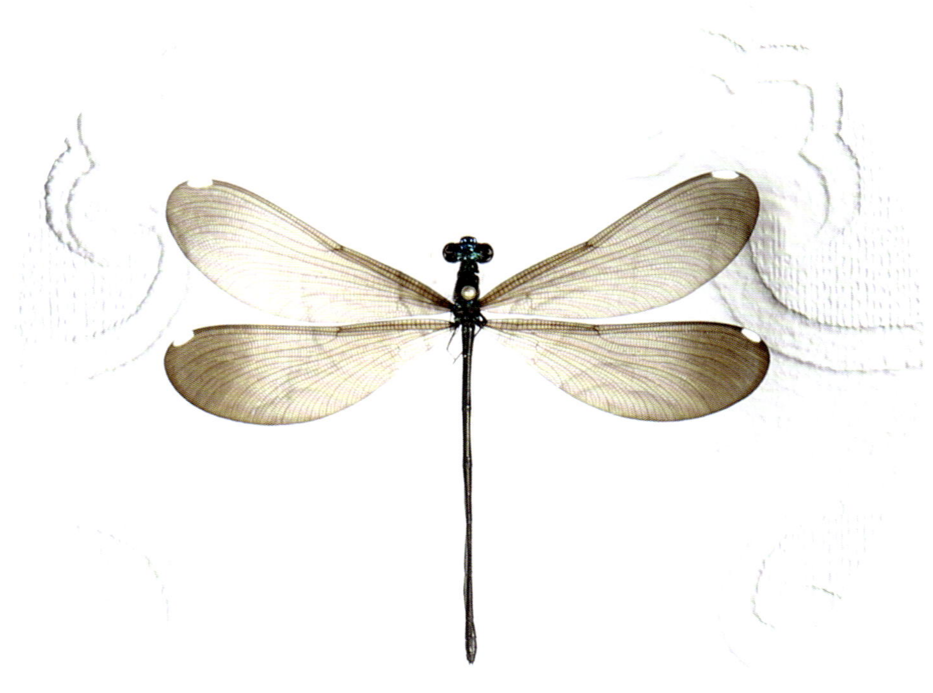

EBONY JEWELWING DAMSELFLY

SCIENTIFIC NAME
Calopteryx maculata

GEOGRAPHIC RANGE
North America

This enchanting insect flutters gracefully around streams in shaded wooded areas, delighting observers with its movements, and it rarely flies very far from its watery habitat. Females in this species crawl partially or fully underwater to carefully deposit their eggs on submerged vegetation, often spending several minutes beneath the surface.

TARANTULA HAWK WASP

SCIENTIFIC NAME
Pepsis grossa

GEOGRAPHIC RANGE
Americas

The tarantula hawk possesses one of the most intense stings of any wasp species, described by many as electrifying. These wasps hunt down, paralyze, and bury one tarantula with each of their eggs. When the egg hatches, the larva begins feeding on the still-living tarantula.

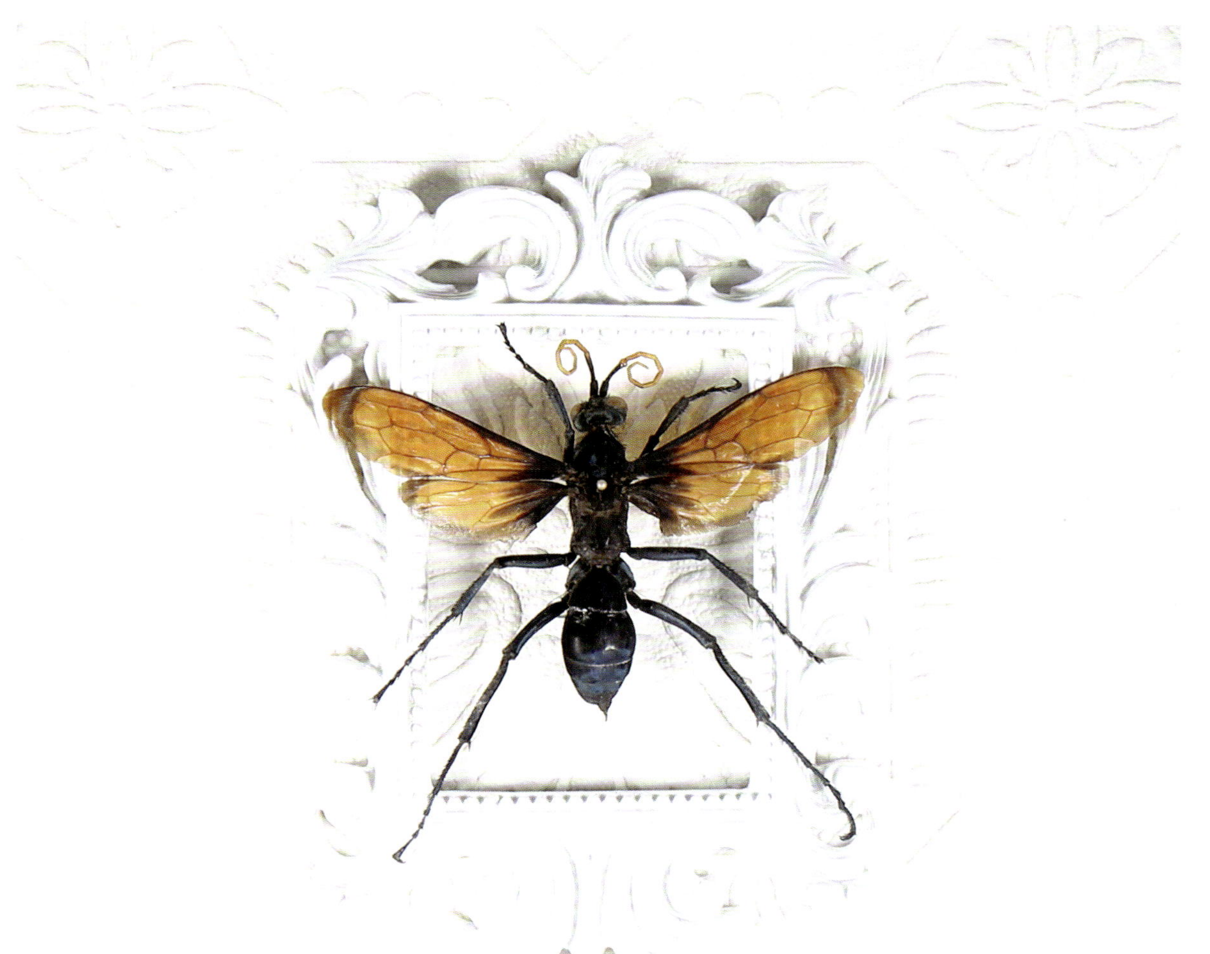

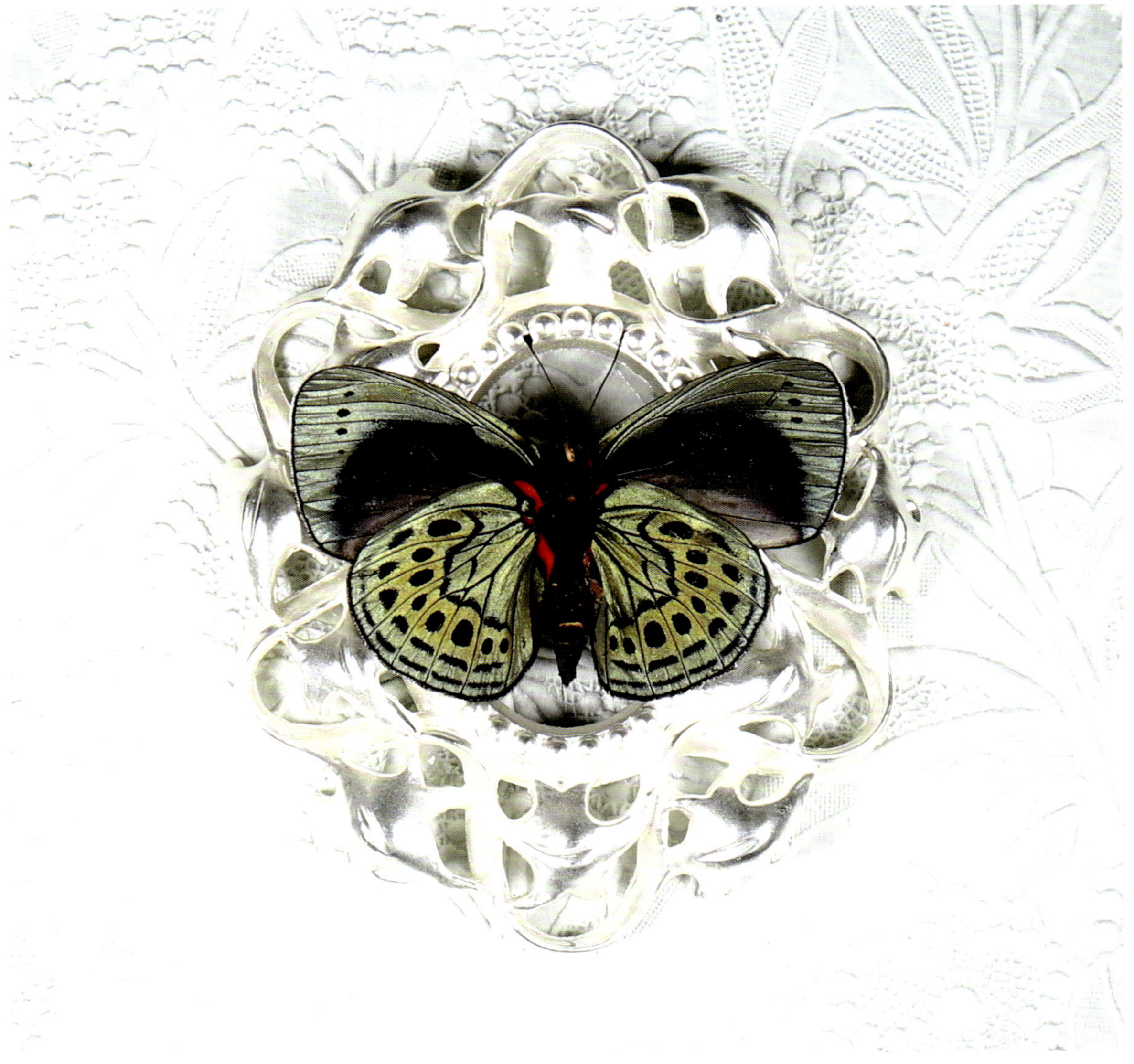

LEPRIEUR'S GLORY

Scientific Name
Asterope leprieuri

Geographic Range
Colombia, Ecuador, Peru,
Bolivia, and Brazil

Populations of this butterfly tend to be highly localized within Amazonia, restricted to areas where their specific host plants thrive. This makes them vulnerable to even small-scale habitat disturbance.

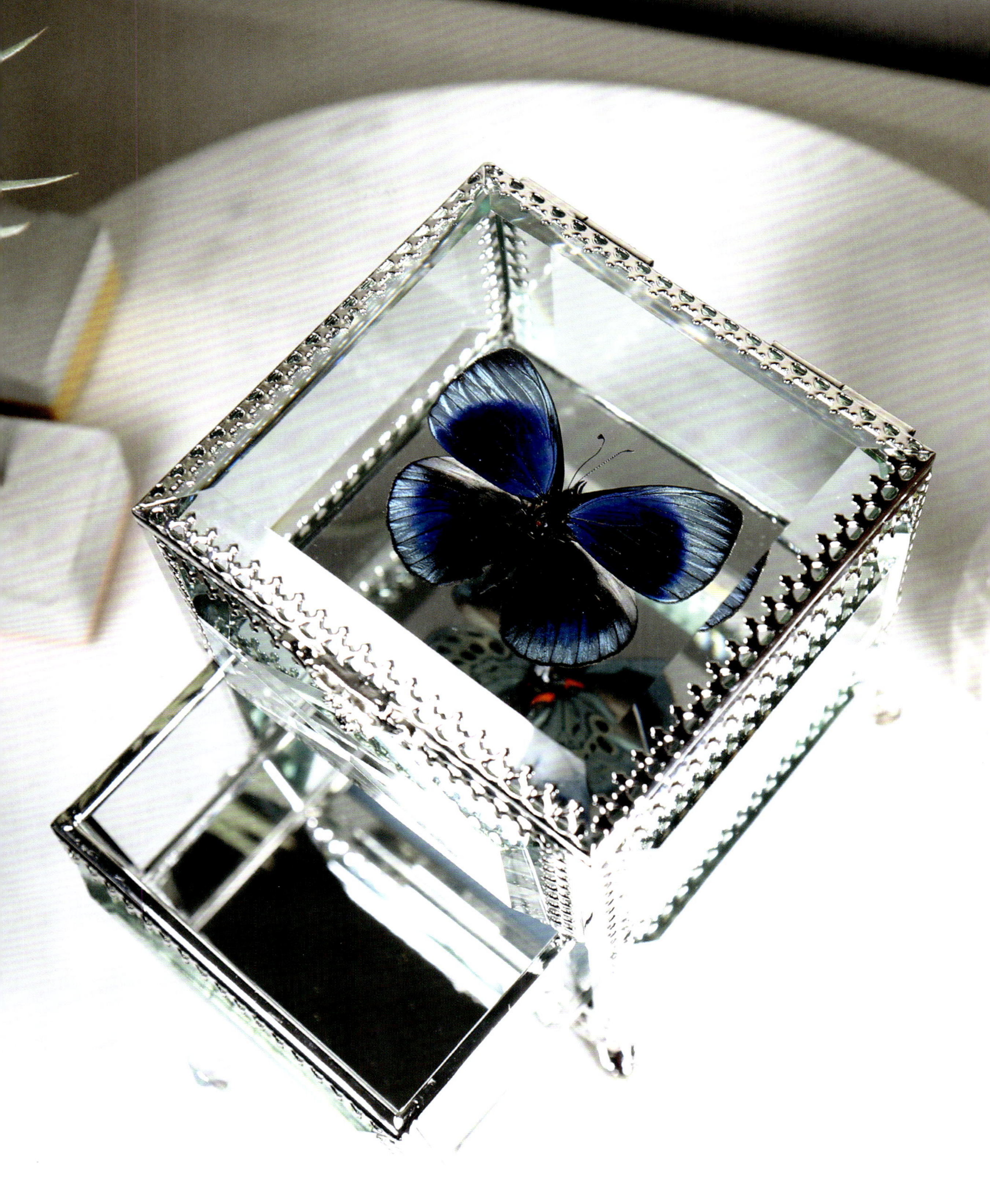

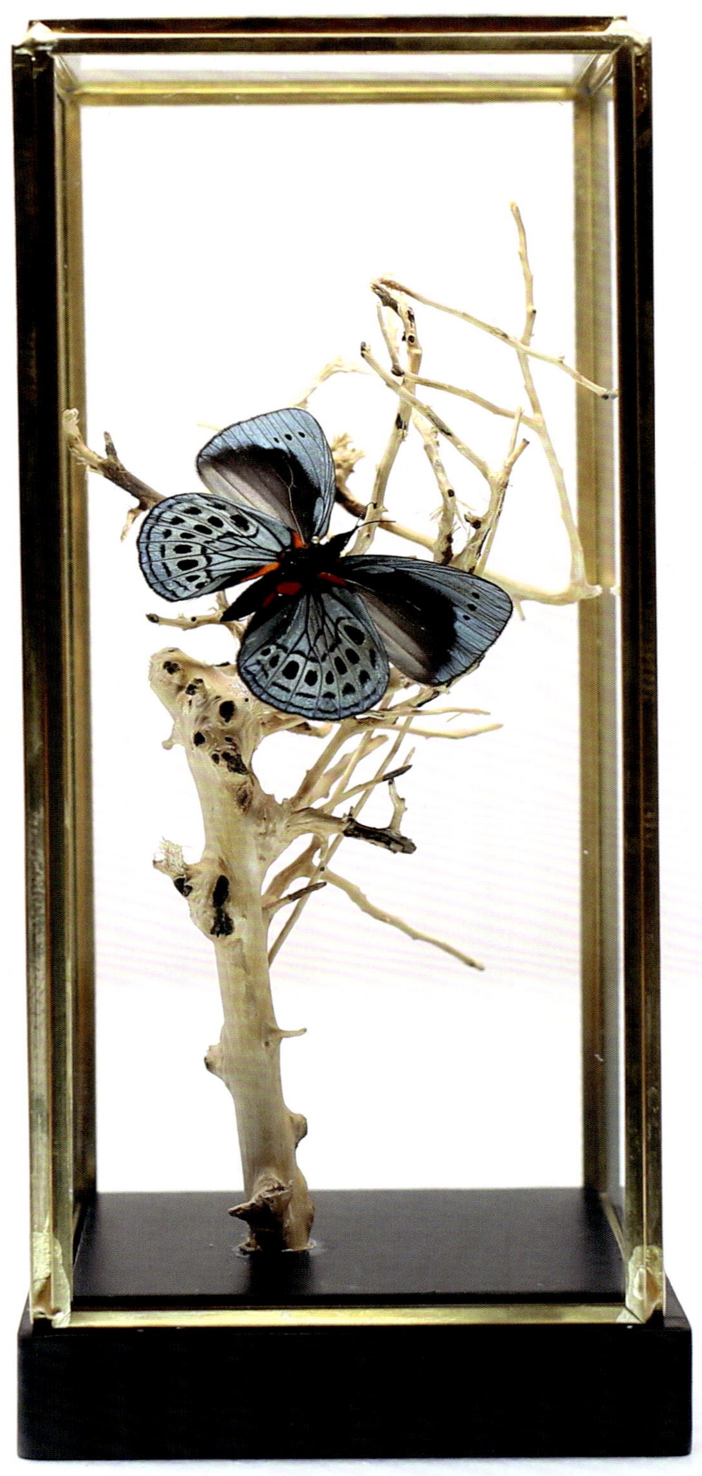

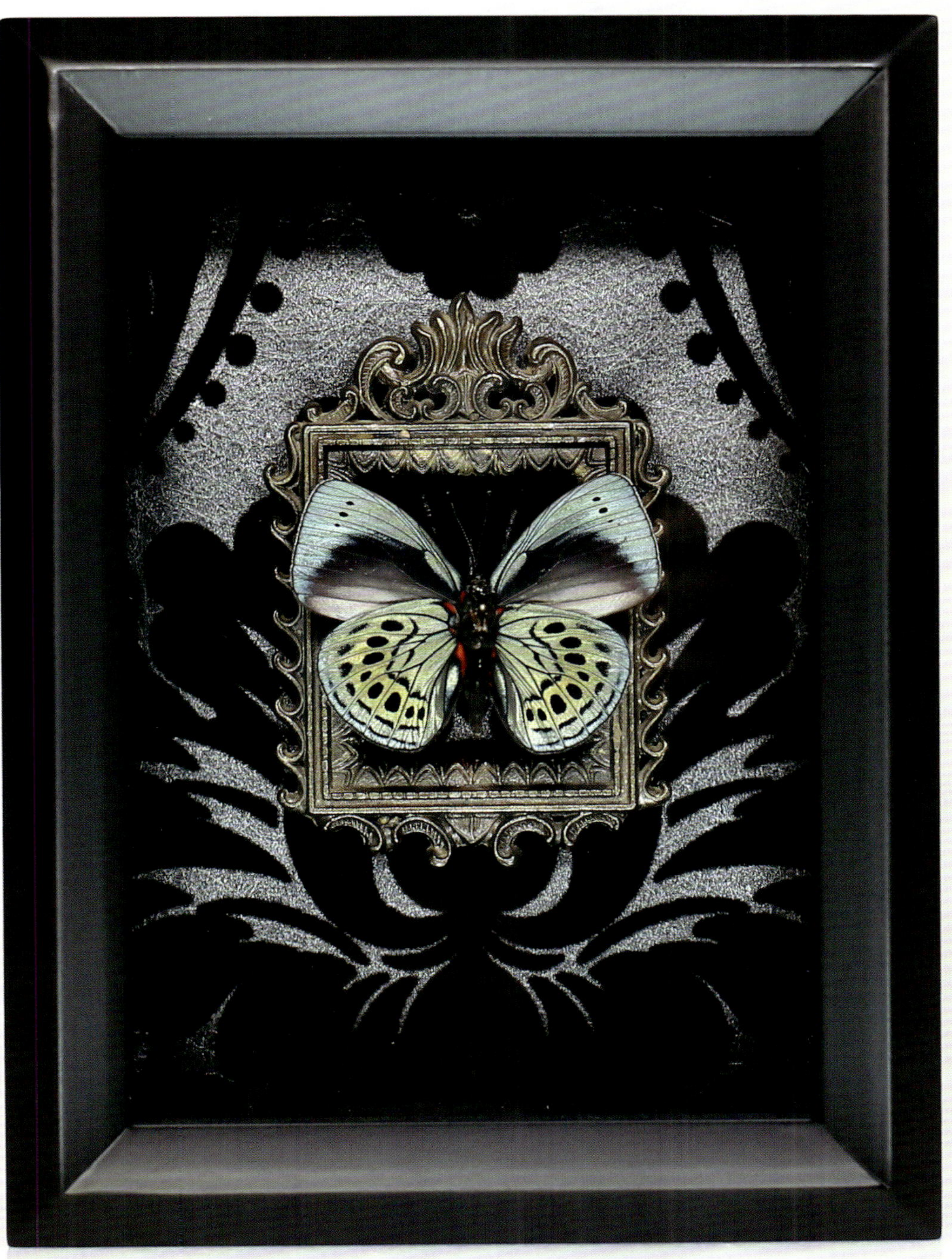

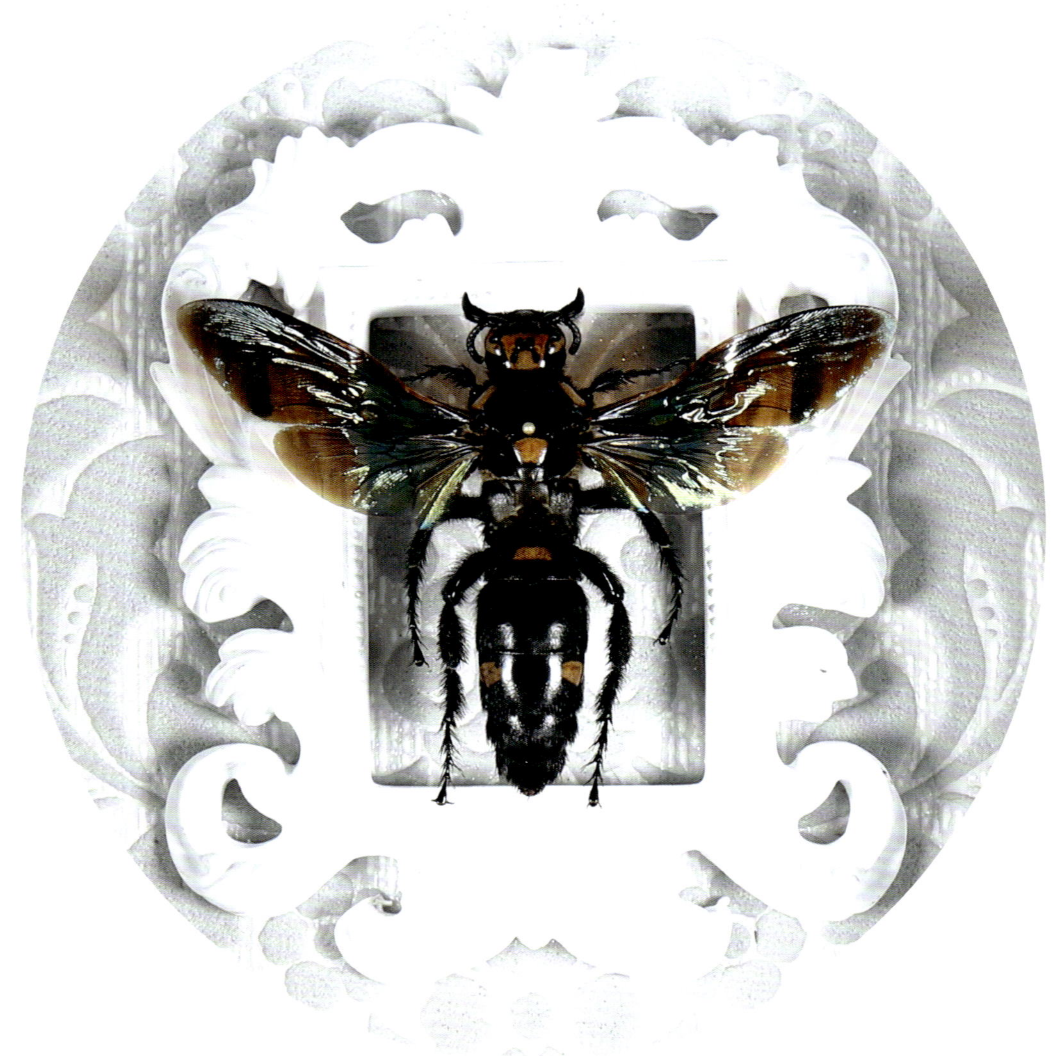

GIANT SCOLIID WASP

SCIENTIFIC NAME
Megascolia procer

GEOGRAPHIC RANGE
India and Southeast Asia

The giant scoliid is one of the largest wasps in the world, boasting an impressive wingspan of 4.6 inches (11.7 centimeters). They specialize in hunting down beetle grubs, which they transport back to their burrows to lay their eggs on. Like the larva of the tarantula hawk moth, the giant scoliid larva eats its host while its host is still alive, preserving vital organs for last to keep the host from rotting.

ATLAS BEETLE

SCIENTIFIC NAME
Chalcosoma atlas

GEOGRAPHIC RANGE
Southeast Asia

Male atlas beetles engage in intense, ritualized combat using their horns to pry, flip, or push rivals off tree trunks. These battles, which occur over access to sap flows or potential mates, rarely cause fatal injuries.

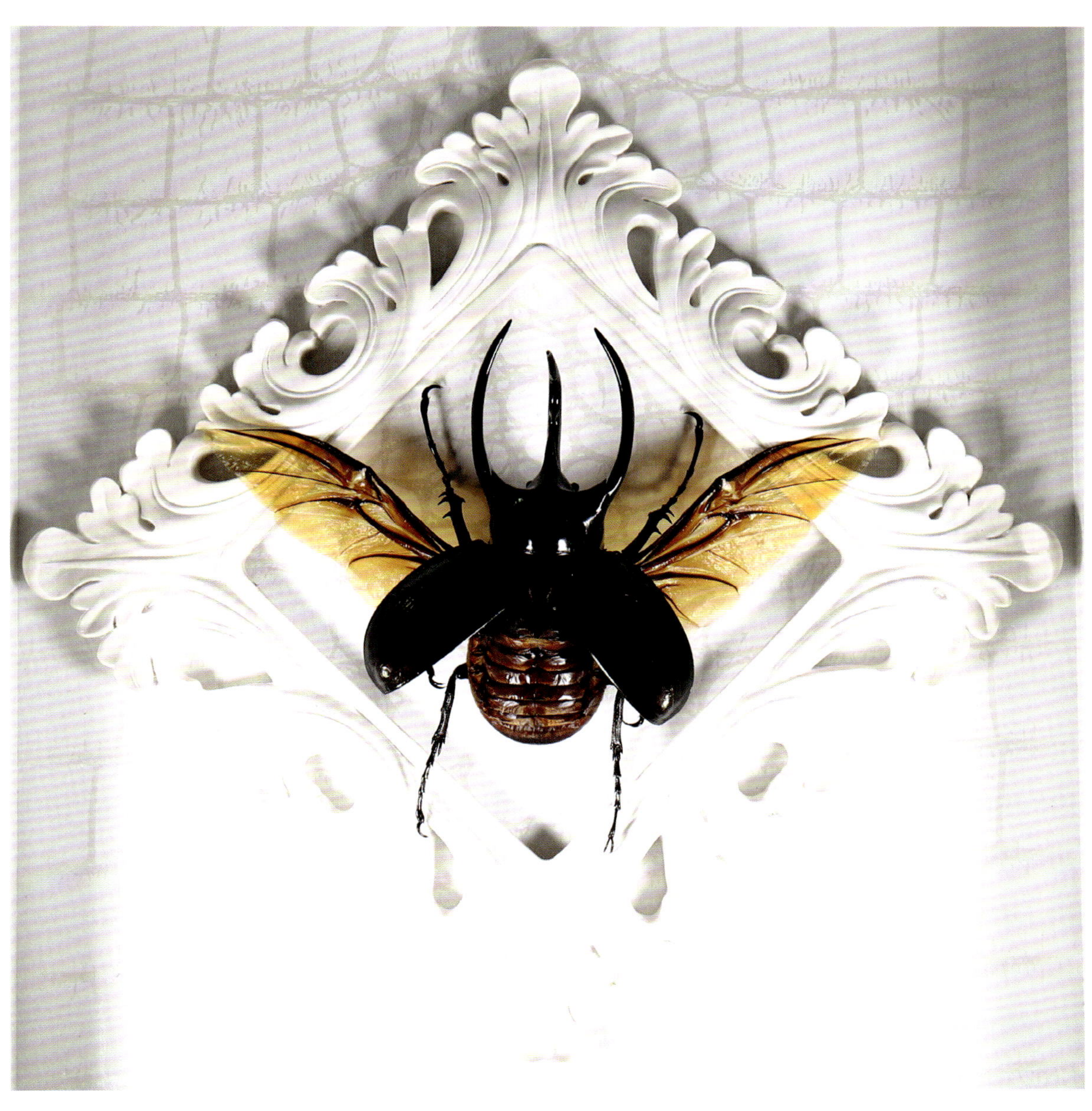

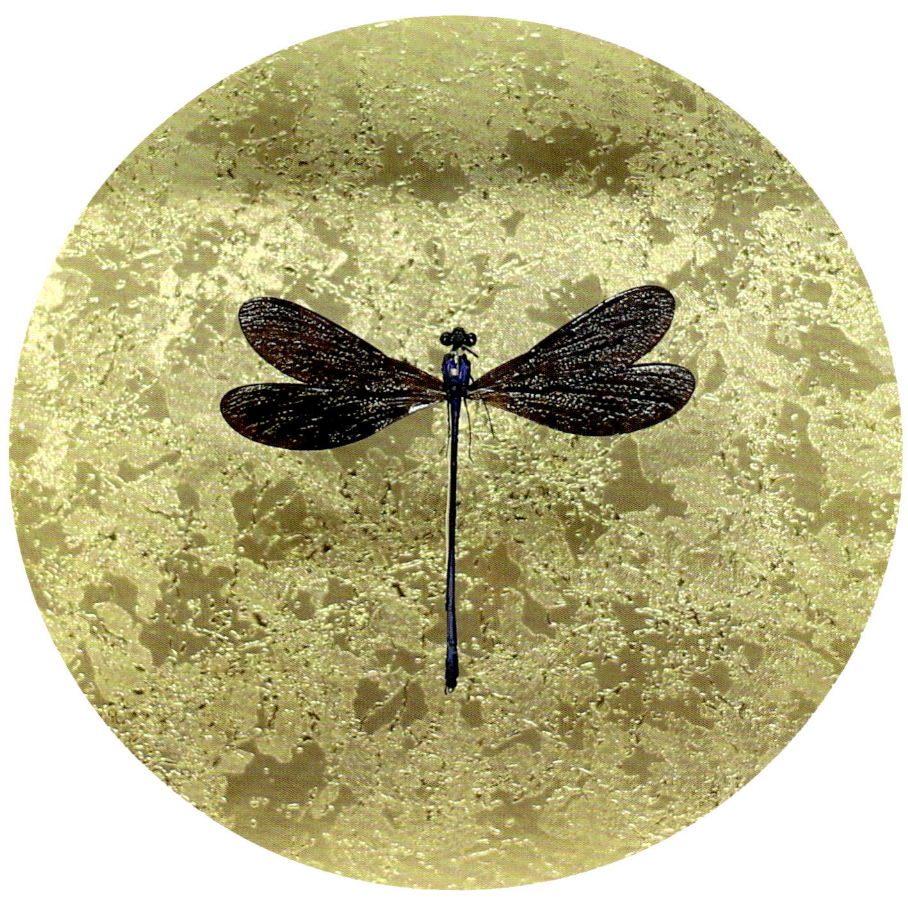

NILA FLASHWING

SCIENTIFIC NAME
Vestalis luctuosa

GEOGRAPHIC RANGE
Indonesia

These creatures primarily inhabit the serene environment of forest streams. The term *Vestalis* refers to the Roman goddess of the hearth, Vesta, possibly alluding to the damselfly's grace. *Luctuosa* means "mournful" in Latin, perhaps referencing the damselfly's dark, somber wing coloration, which evokes a sense of melancholy.

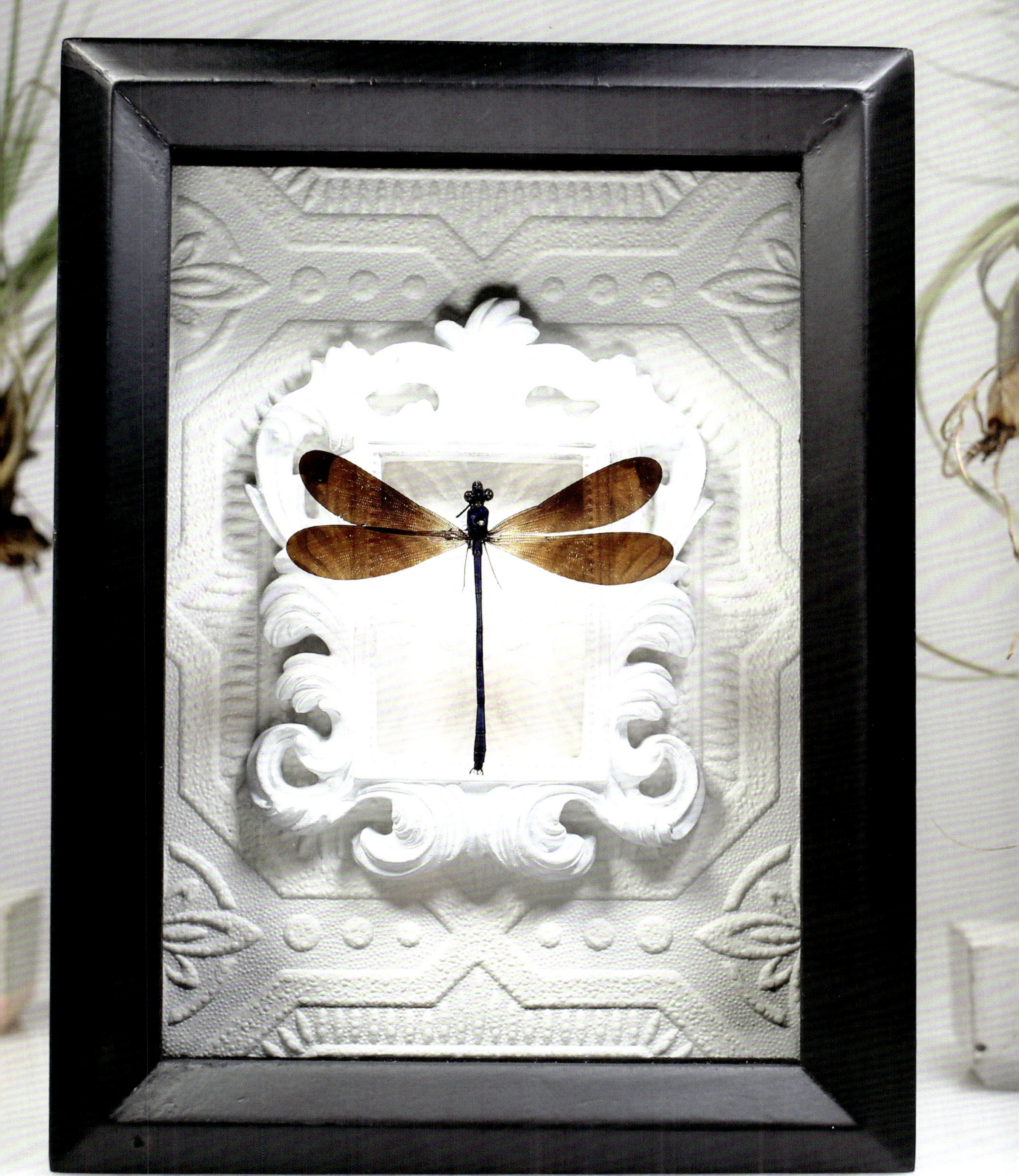

THE TACTILE WORLD

TEXTURE IN INSECTS

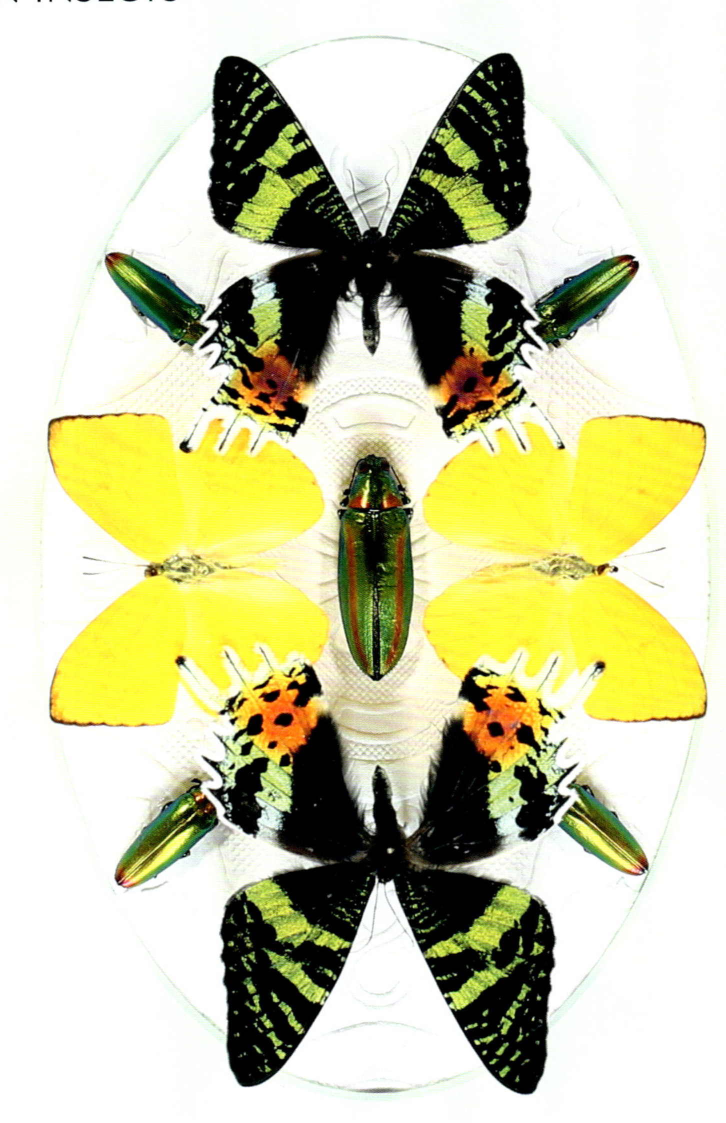

INSECT BODIES ARE OFTEN ADORNED WITH TEXTURED SURFACES, RANGING from the velvety softness of a moth's wings to the spiny, armored exoskeleton of a sturdy beetle. These diverse textures, though sometimes subtle and overlooked by the human eye, play a central role in shaping how insects interact with their varied environments. They influence how insects move, protect themselves, and communicate in intricate ways we are only beginning to understand.

The staggering variety of textures found in insects is nothing short of incredible: fine, silky hairs that enable them to detect even the slightest air movement; bristles that assist in deft navigation through dense undergrowth; and scales that provide excellent camouflage or facilitate effortless flight.

Insects have become masters of the art of tactile adaptation. For example, the rough, bumpy surface of a beetle's sturdy shell might serve to deter a hungry predator, while the soft, smooth body of a caterpillar may help it blend into the leaves it eagerly feeds on. The hairlike structures found on a moth's wings and antennae can pick up the faintest vibrations, allowing it to skillfully navigate the subtle movements of the air or even detect the presence of a mate from great distances. Other structures are specifically designed for aggression or defense—such as the prickly body of an intimidating hedgehog bark beetle—offering powerful deterrents to any predator that dares to strike. Ultimately, texture in the insect kingdom is as varied as it is vital, with each surface telling its own story of evolution, adaptation, and survival.

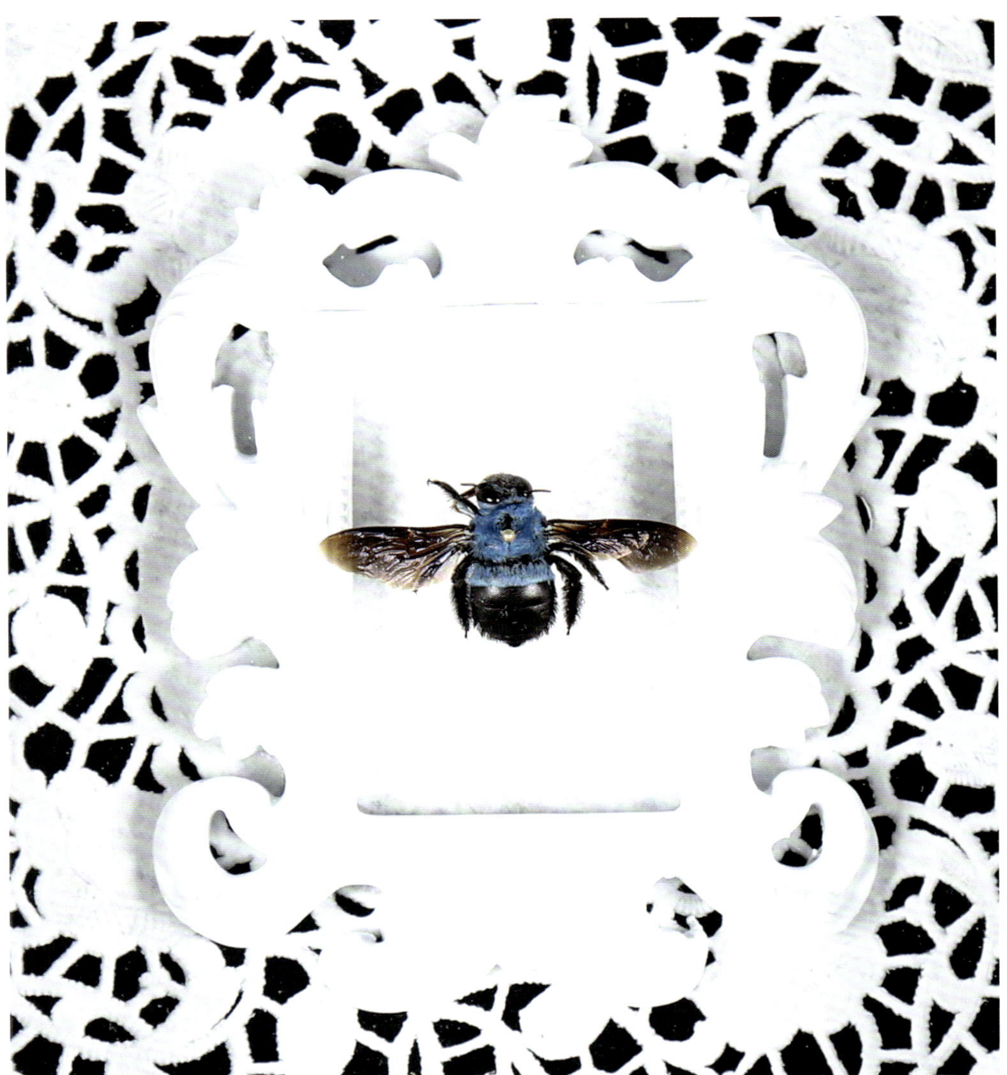

BLUE CARPENTER BEE

SCIENTIFIC NAME
Xylocopa caerulea

GEOGRAPHIC RANGE
India, China, and
Southeast Asia

Among blue carpenter bees, it's the females who shine. Their bodies shimmer with a vivid blue sheen, making them easy to spot, especially compared with the more muted, dusty-brown males. These bees live alone rather than in colonies, tunneling into tree trunks to build their nests and lay eggs. While they do produce honey, it's not like the familiar golden honey of European honeybees—it's thick and concentrated, closer in texture to a dense paste.

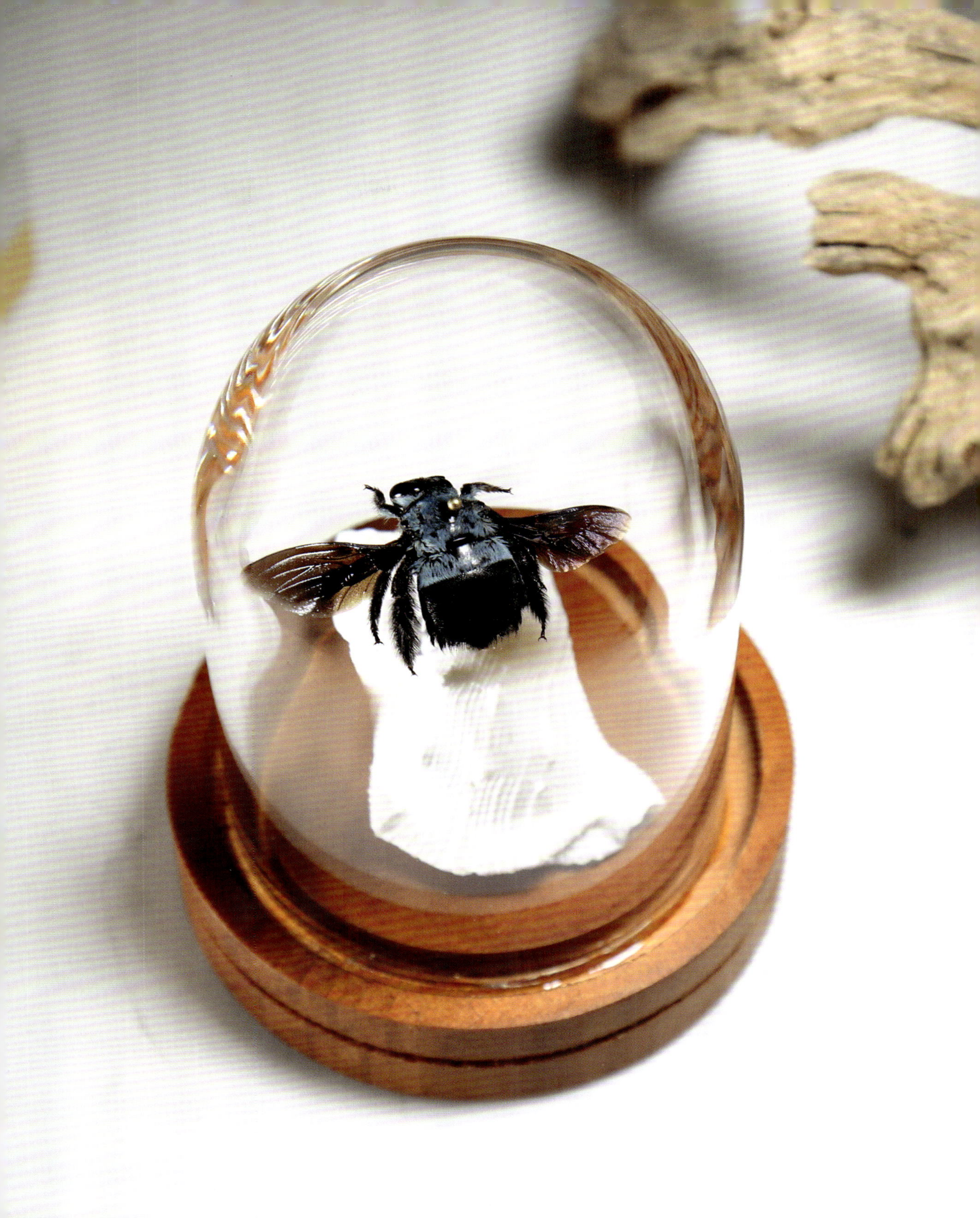

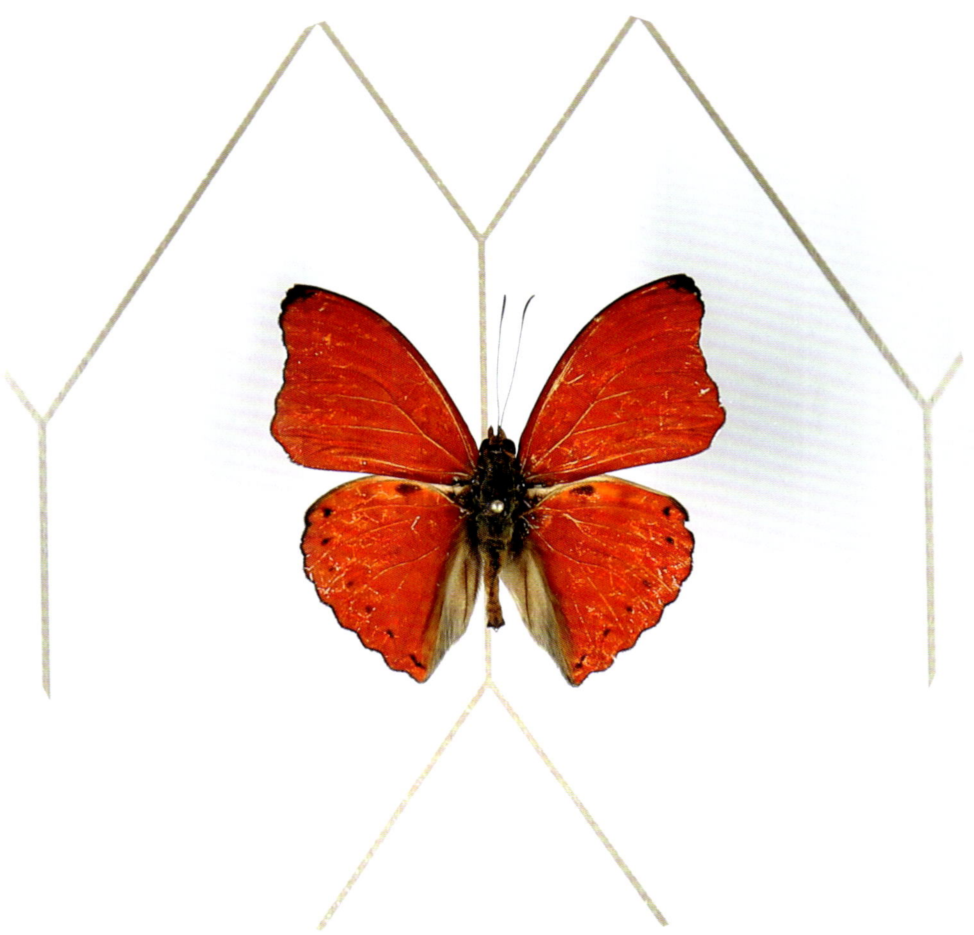

BLOOD-RED GLIDER

SCIENTIFIC NAME
Cymothoe sangaris

GEOGRAPHIC RANGE
Central Africa

The pupa of this species mimics a curled, dead leaf hanging from the underside of foliage—a classic example of morphological crypsis, or camouflage, that protects it from visual hunters like birds.

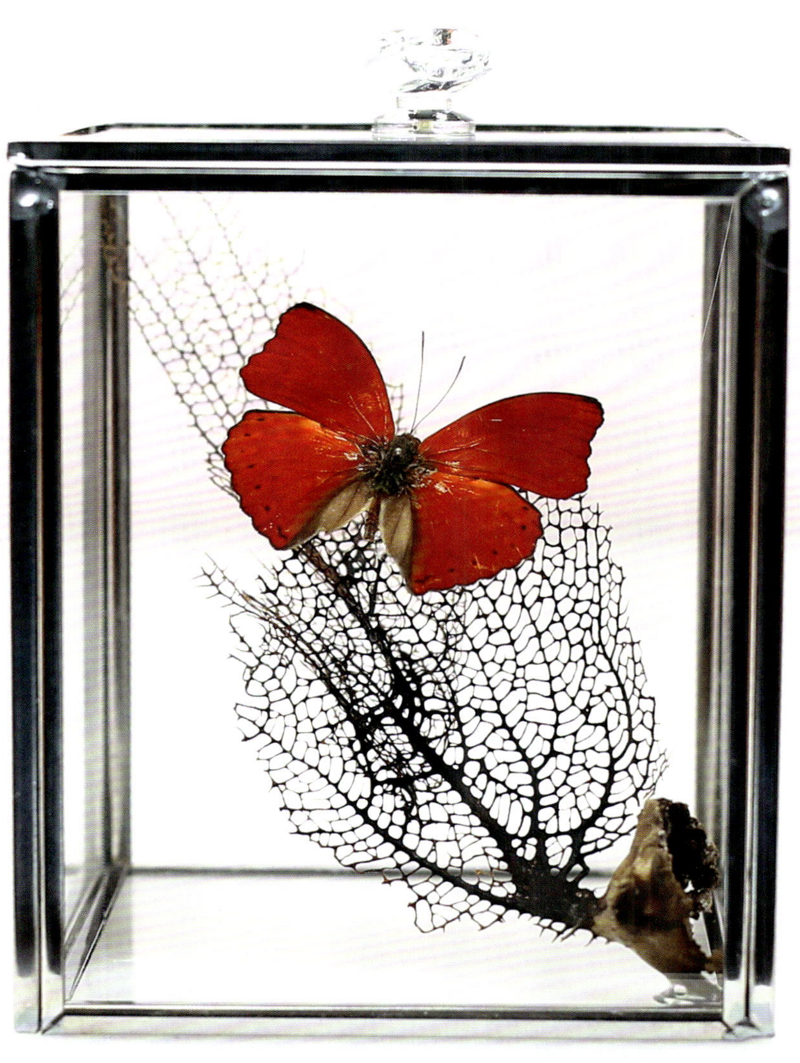

ELEPHANT DUNG BEETLE

Female dung beetles construct enormous dung balls (up to the size of an orange) as shelter for a single egg. They bury the dung with the egg inside, and it provides nutrition, moisture regulation, and protection until the beetle emerges as an adult.

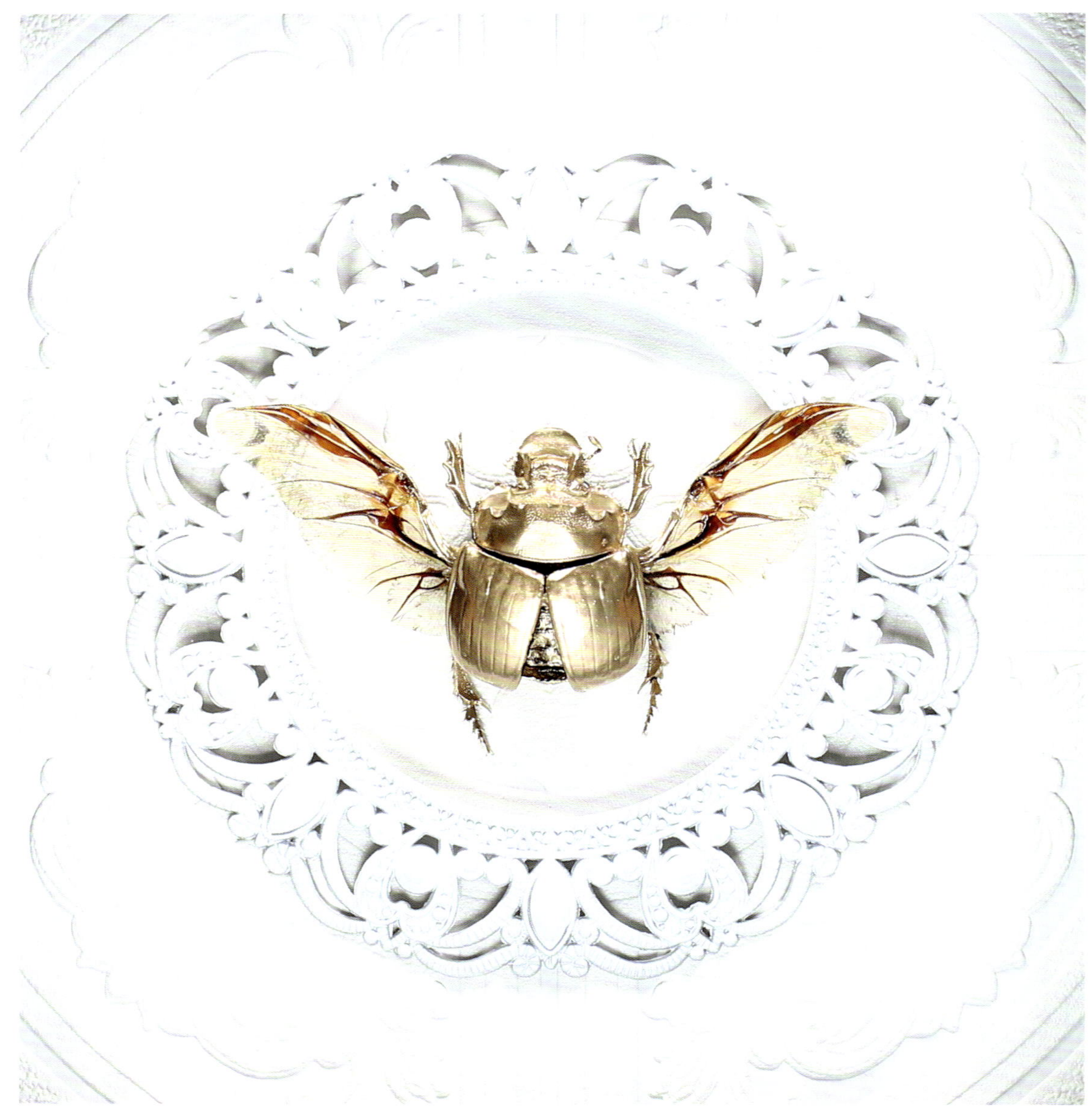

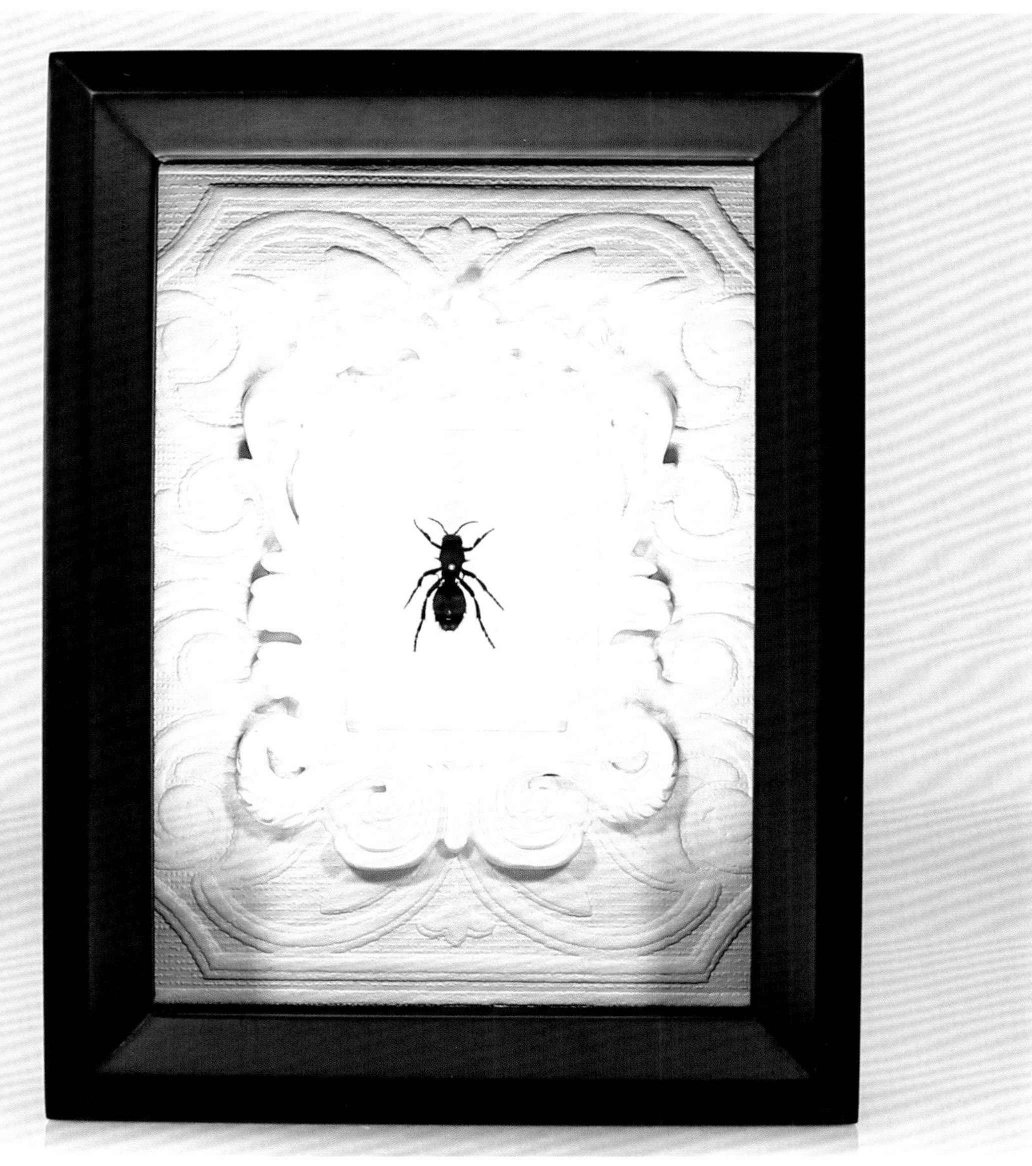

EASTERN VELVET ANT

SCIENTIFIC NAME
Dasymutilla occidentalis

GEOGRAPHIC RANGE
Central and Eastern
United States

The velvet ant, which is commonly referred to as the cow killer because some believe its sting is strong enough to kill a cow, is actually a type of wasp. The females of this species are wingless and bear a strong resemblance to ants, leading to their misnomer.

GIANT RED-WINGED GRASSHOPPER

SCIENTIFIC NAME
Tropidacris cristata dux

GEOGRAPHIC RANGE
Mexico and Central America

Unlike adults, nymphs in this species lack fully developed wings and cannot fly. Instead, they use their exceptionally strong hind legs to make long, coordinated jumps when disturbed, sometimes launching themselves several body lengths in a single bound.

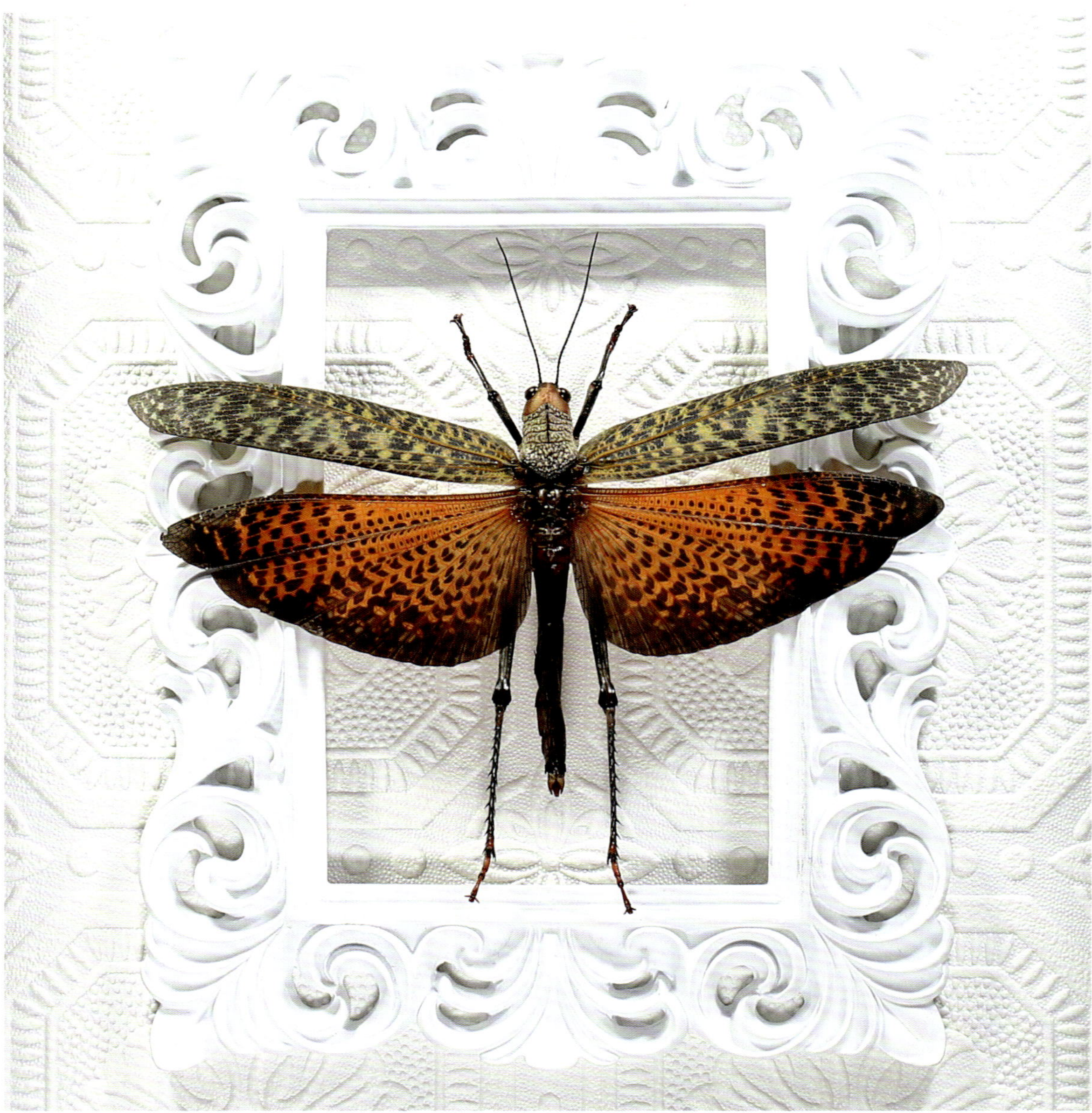

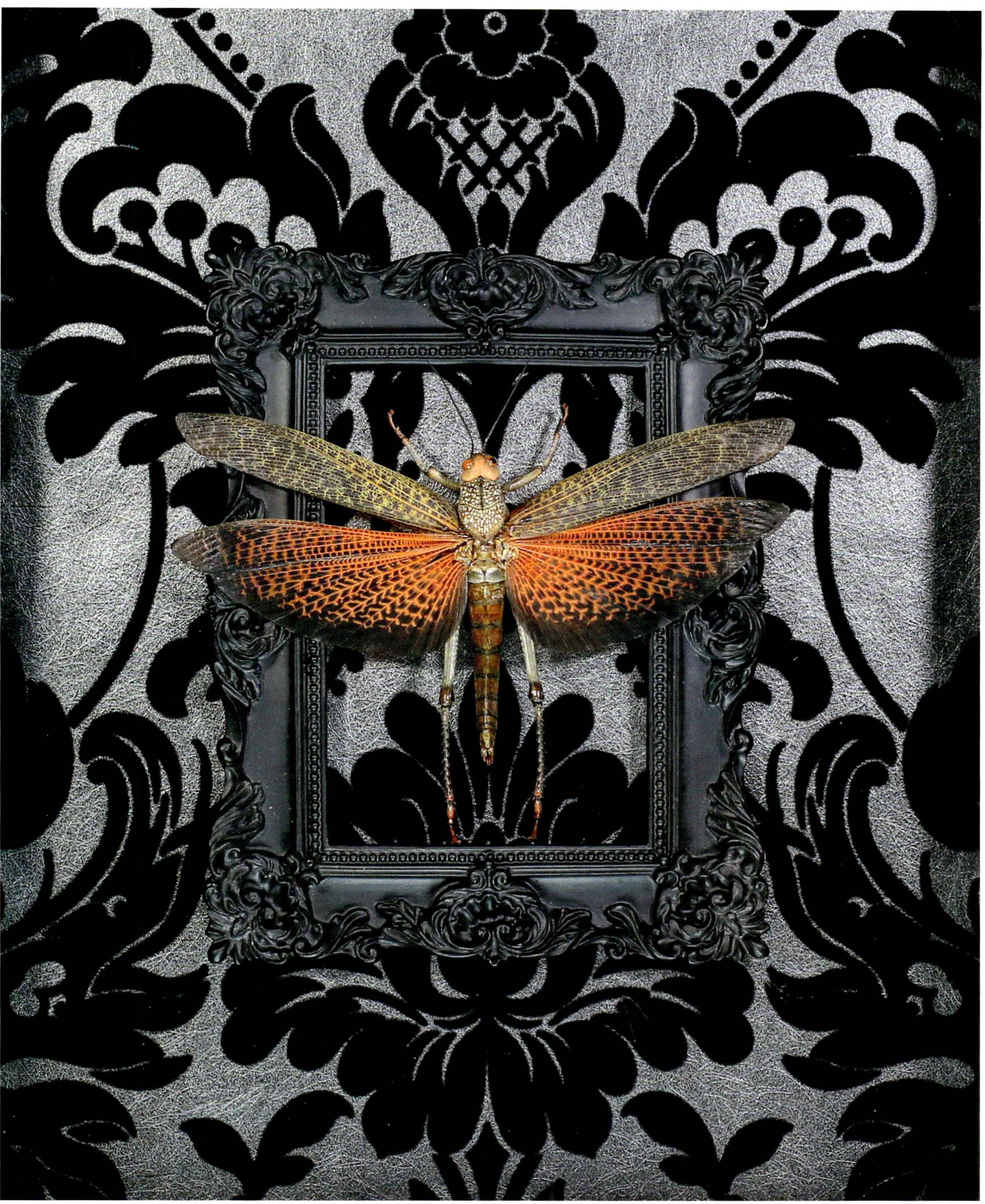

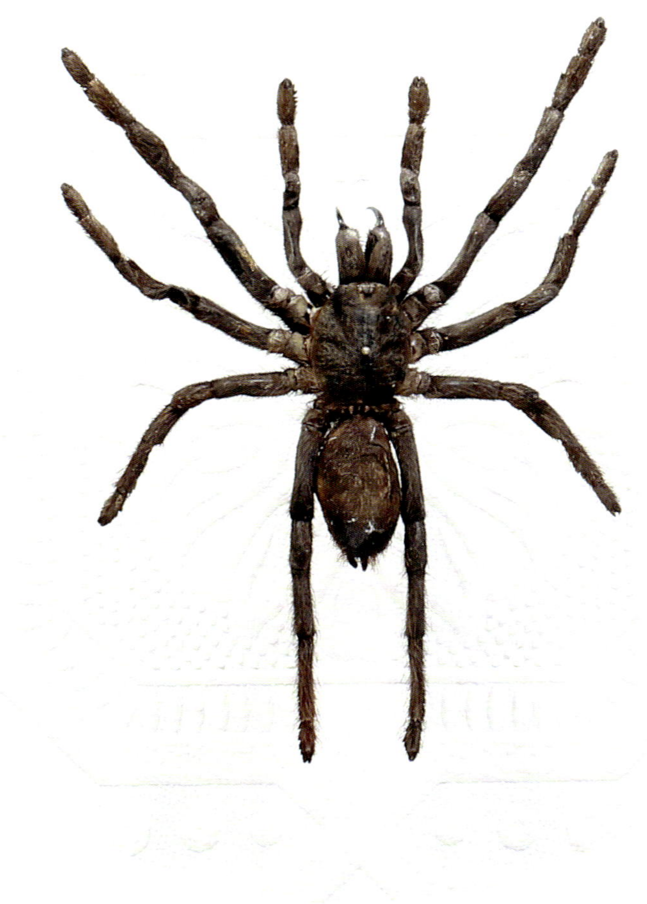

CHILOBRACHYS NITELINUS TARANTULA

SCIENTIFIC NAME
Chilobrachys nitelinus

GEOGRAPHIC RANGE
Sri Lanka

Chilobrachys tarantulas use their jawlike chelicerae organs, which are composed of a series of short, fine spines, to produce a distinctive clicking sound. This behavior, known as stridulation, is used primarily for communication or as a defense mechanism.

EUROPEAN HONEYBEE

Scientific Name
Apis mellifera

Geographic Range
Global, except Antarctica

The European honeybee's genome was sequenced to gain a deeper understanding of their intricate social behavior, their complex evolution, and the underlying genetic basis for various traits that are important for overall bee health and effective pollination processes. These amazing creatures play a crucial role in the pollination of essential food crops, and if the bees were to disappear entirely, their absence would inevitably result in a significant agricultural decline impacting global populations.

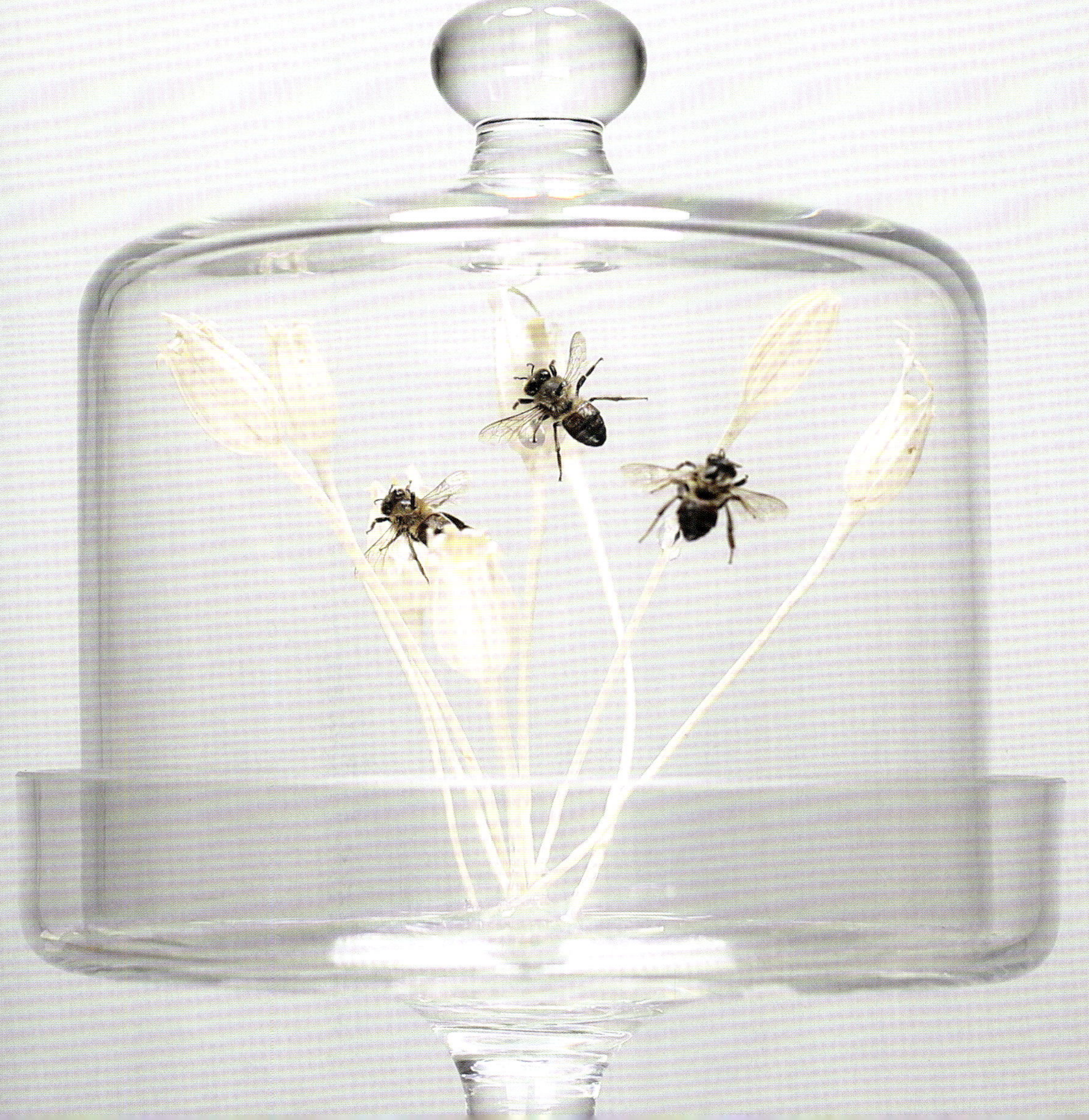

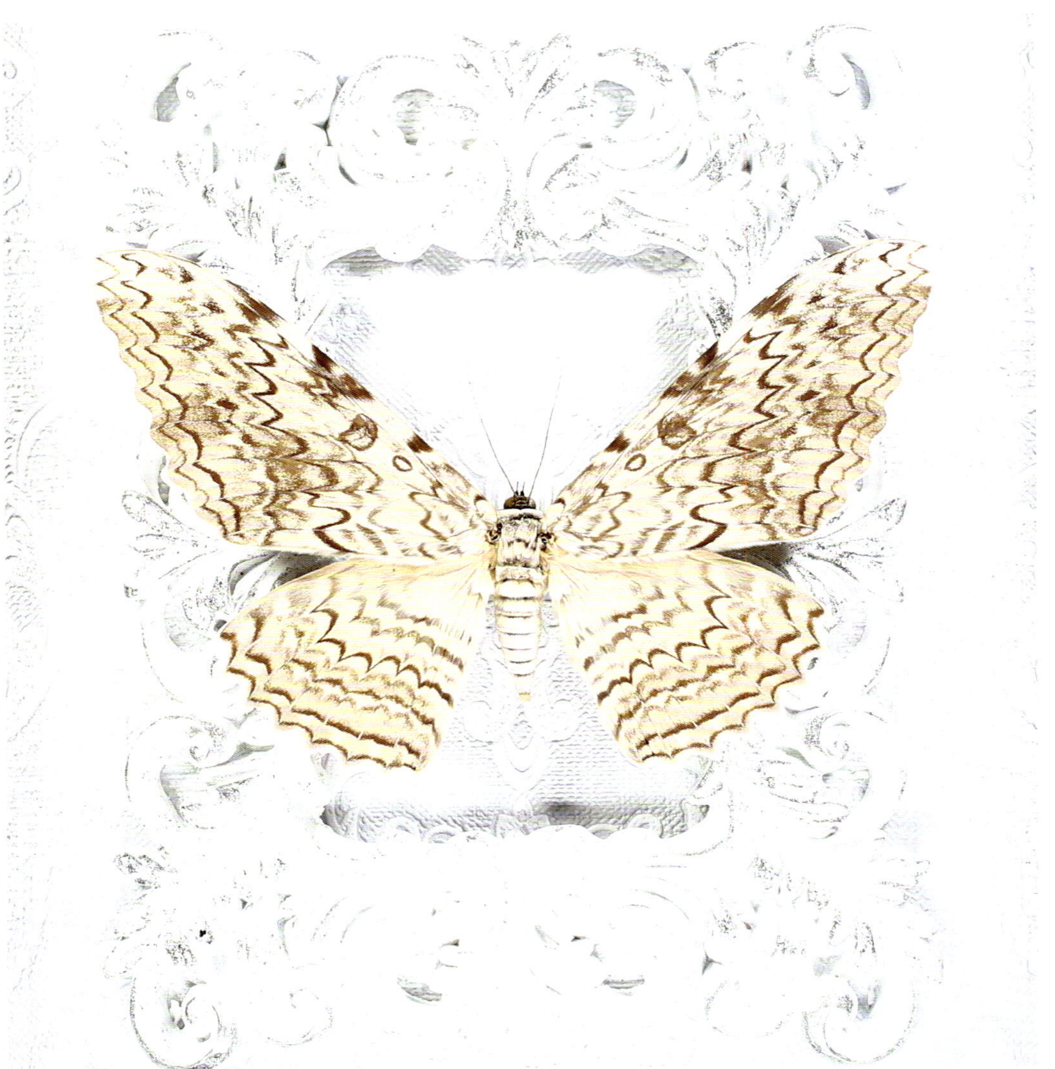

WHITE WITCH MOTH

SCIENTIFIC NAME
Thysania agrippina

GEOGRAPHIC RANGE
Mexico, Central America,
and South America

This extraordinary moth species holds the record for the largest recorded wingspan of any living insect, with reports of wingspans up to 12 inches (30.5 centimeters). Its common name, white witch moth, may stem from the experiences of early naturalists who found it nearly impossible to shoot down the species. The name lends an air of mystery to the moth's already striking presence in nature.

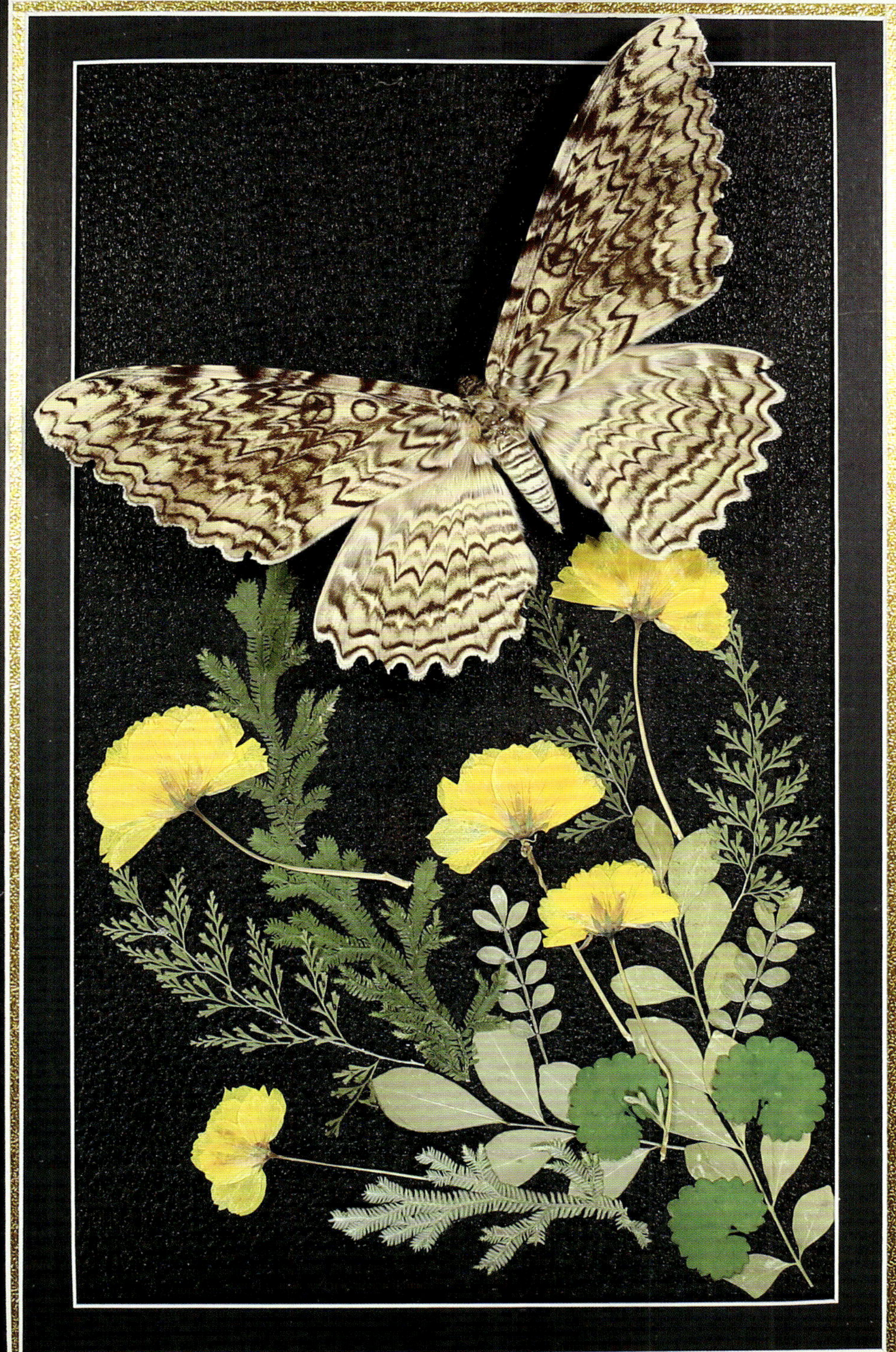

BLUE-WINGED GRASSHOPPER

SCIENTIFIC NAME
Oedipoda caerulescens

GEOGRAPHIC RANGE
Europe, North Africa,
and Asia

At rest, the blue-winged grasshopper is nearly invisible against dry, stony terrain, thanks to its highly effective background matching. But when startled, it bursts into flight with a flash of bright blue hind wings, then drops suddenly to the ground and stays still.

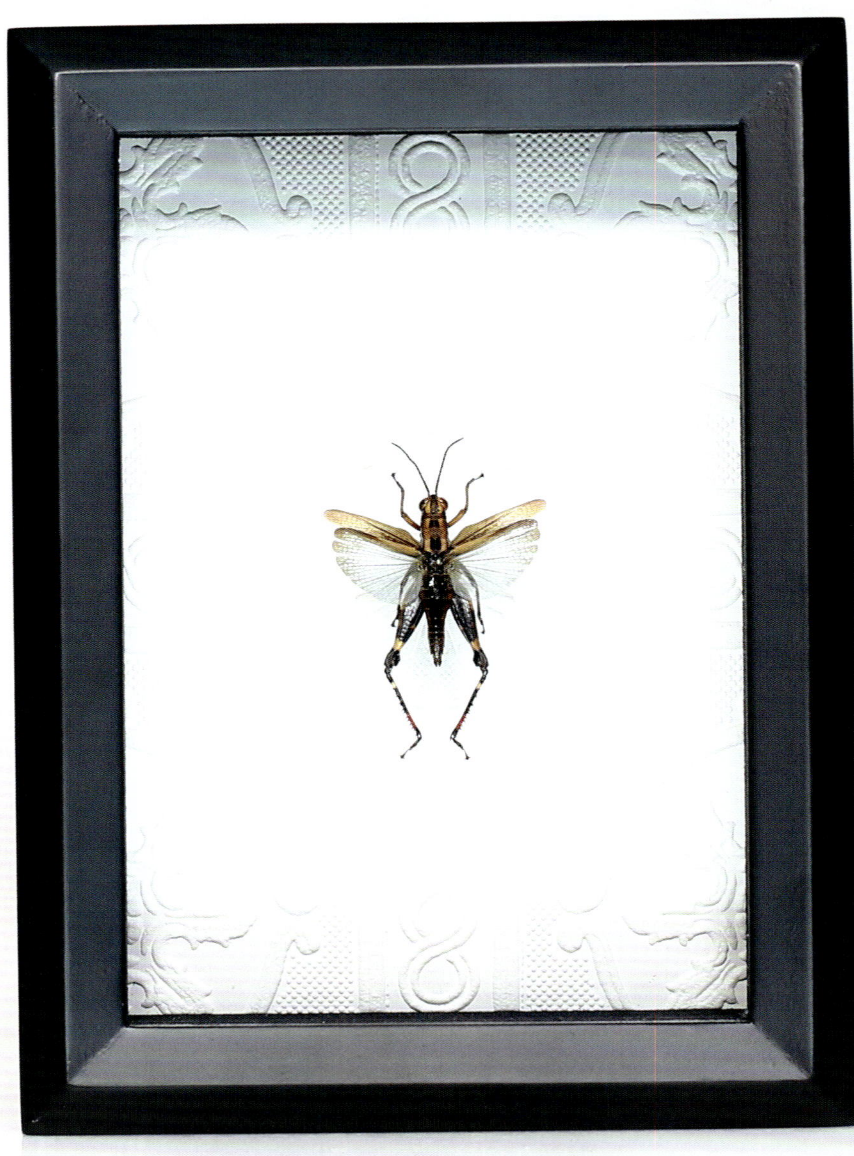

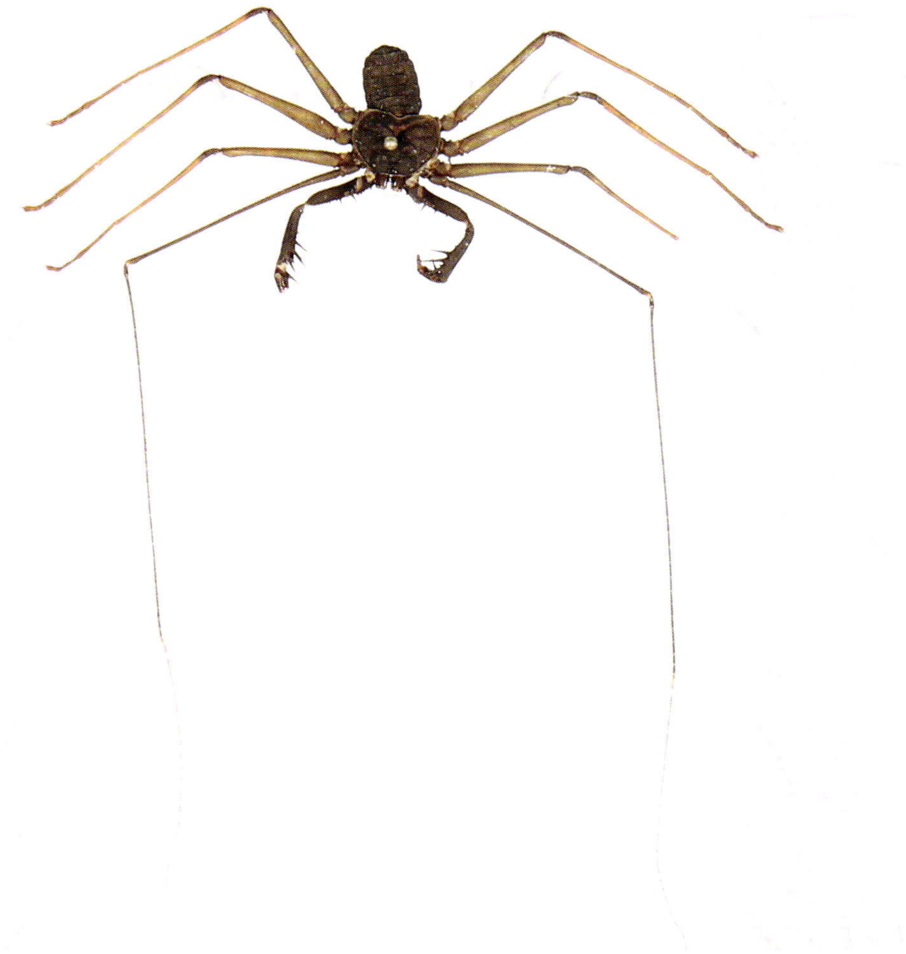

BAJA WHIPSPIDER

SCIENTIFIC NAME
Acanthophrynus coronatus

GEOGRAPHIC RANGE
Mexico

These creatures are not venomous and lack any form of sting. They are nocturnal arachnids that utilize their long, highly sensitive legs to skillfully navigate their environment and hunt for prey. The whipspider is the intriguing creature used to demonstrate the "unforgivable curses" in the movie *Harry Potter and the Goblet of Fire.*

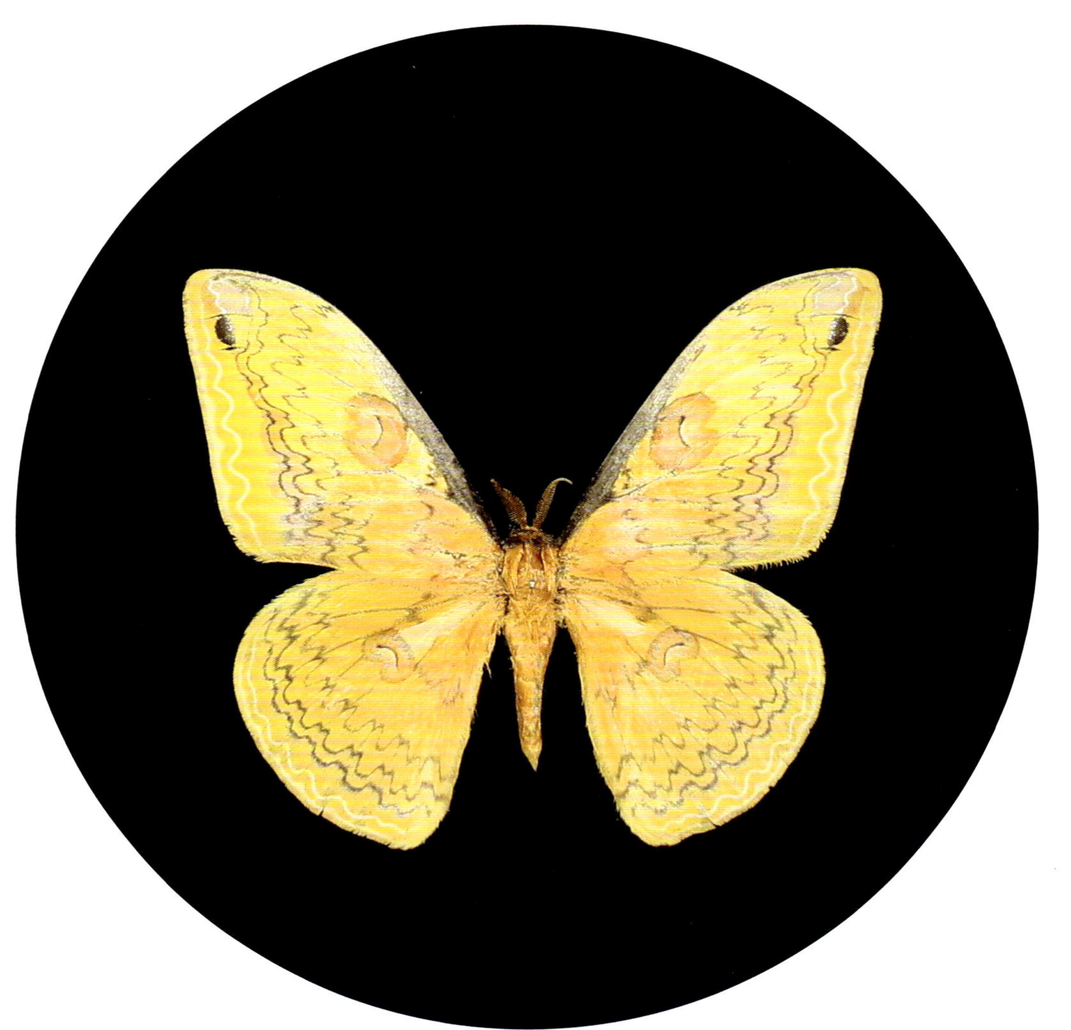

OBERTHÜR'S SILKMOTH

SCIENTIFIC NAME
Loepa oberthuri

GEOGRAPHIC RANGE
China

These moths emerge from strong silk cocoons they've built to protect themselves during their vital transformation into adults. One of their most dramatic features is their distinctive pink-and-black-striped feet.

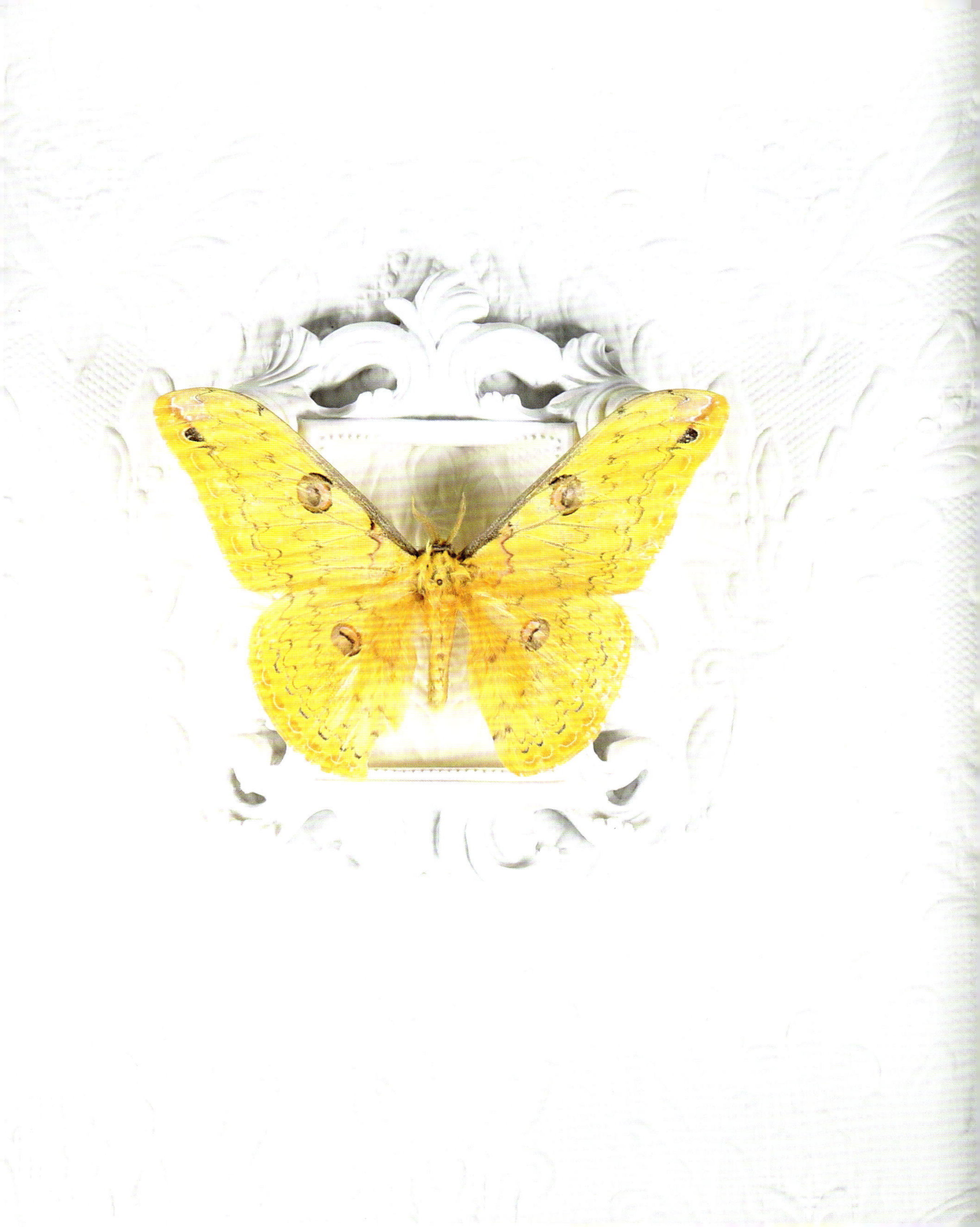

SPINY LOBSTER KATYDID

SCIENTIFIC NAME
Panoploscelis specularis

GEOGRAPHIC RANGE
Guyana, Colombia, Ecuador, Peru, Brazil

These large katydids are terrestrial and somewhat predatory in nature, consuming leaves and preying upon small insects that share their habitat. Even though they possess two pairs of wings, they are classified as brachypterous insects, which means that they are flightless because of their underdeveloped, or vestigial, hind wings.

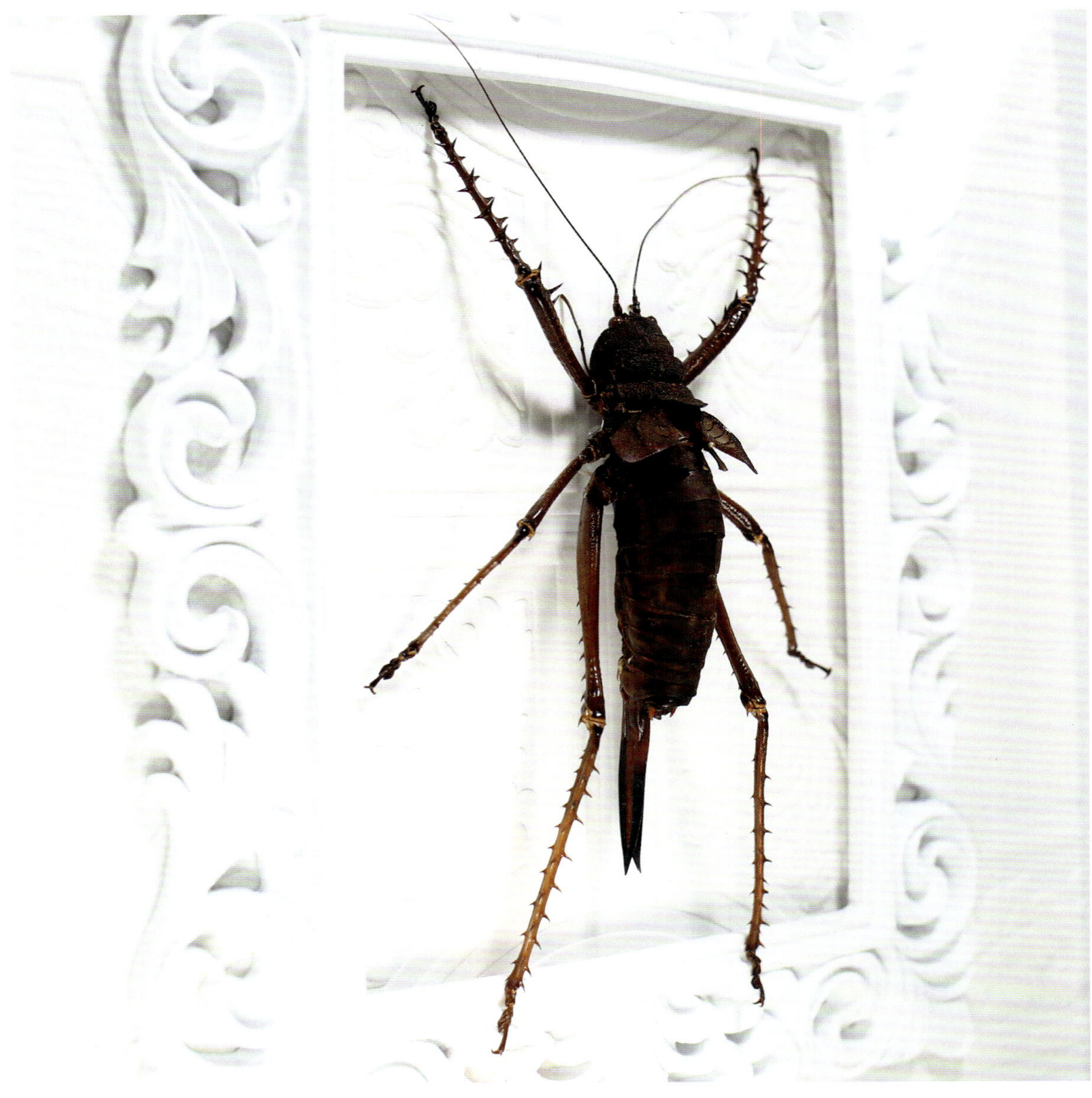

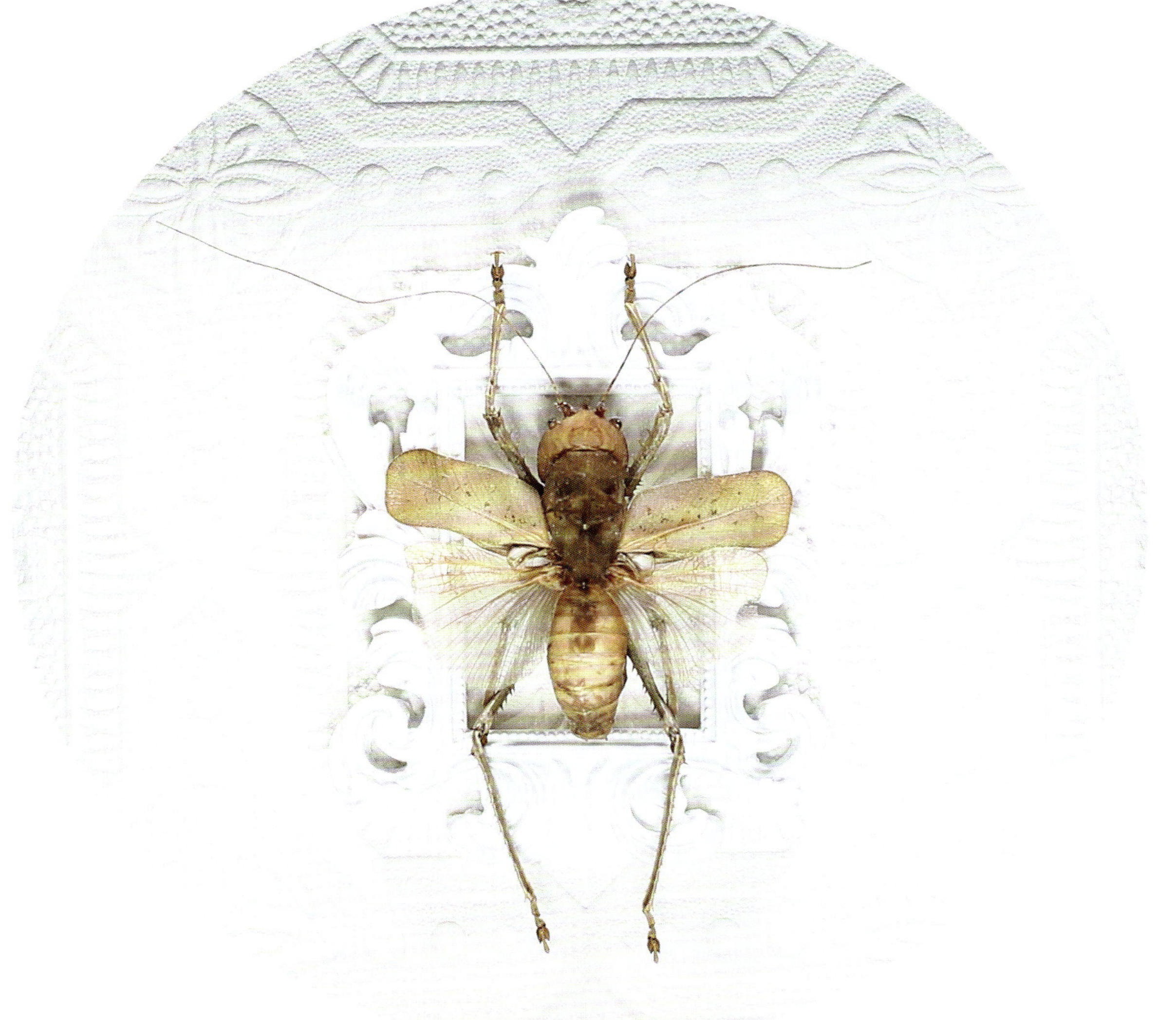

DRAGON-HEADED KATYDID

SCIENTIFIC NAME
Lesina spp.

GEOGRAPHIC RANGE
Malaysia, Brunei,
and Indonesia

Katydids are exceptionally well camouflaged, mimicking the appearance of leaves to avoid detection. Some male dragon-headed katydids produce calls that closely resemble the sound of a rattlesnake's rattle, helping them attract mates and defend their territory.

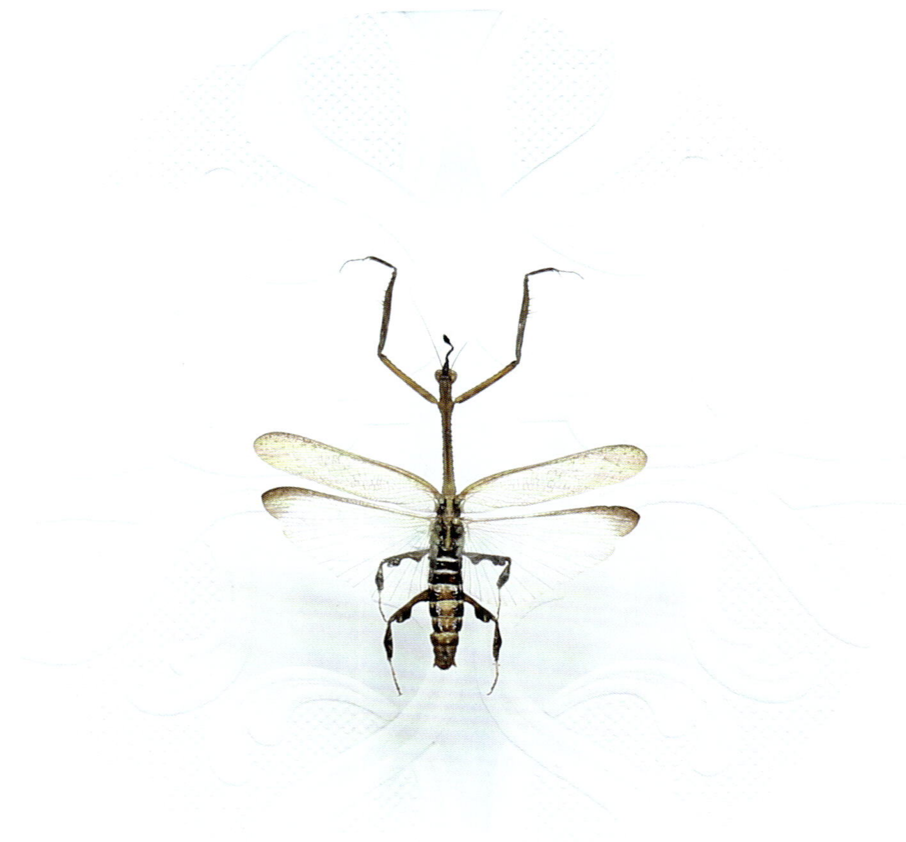

GHOST MANTIS

Scientific Name
Phyllocrania paradoxa

Geographic Range
Africa

The ghost mantis closely resembles a dry, weathered leaf in its natural habitat. When threatened, older nymphs and adult females will often resort to "playing dead," which can be quite effective in deterring potential predators. By contrast, adult males usually run quickly or take flight to escape danger.

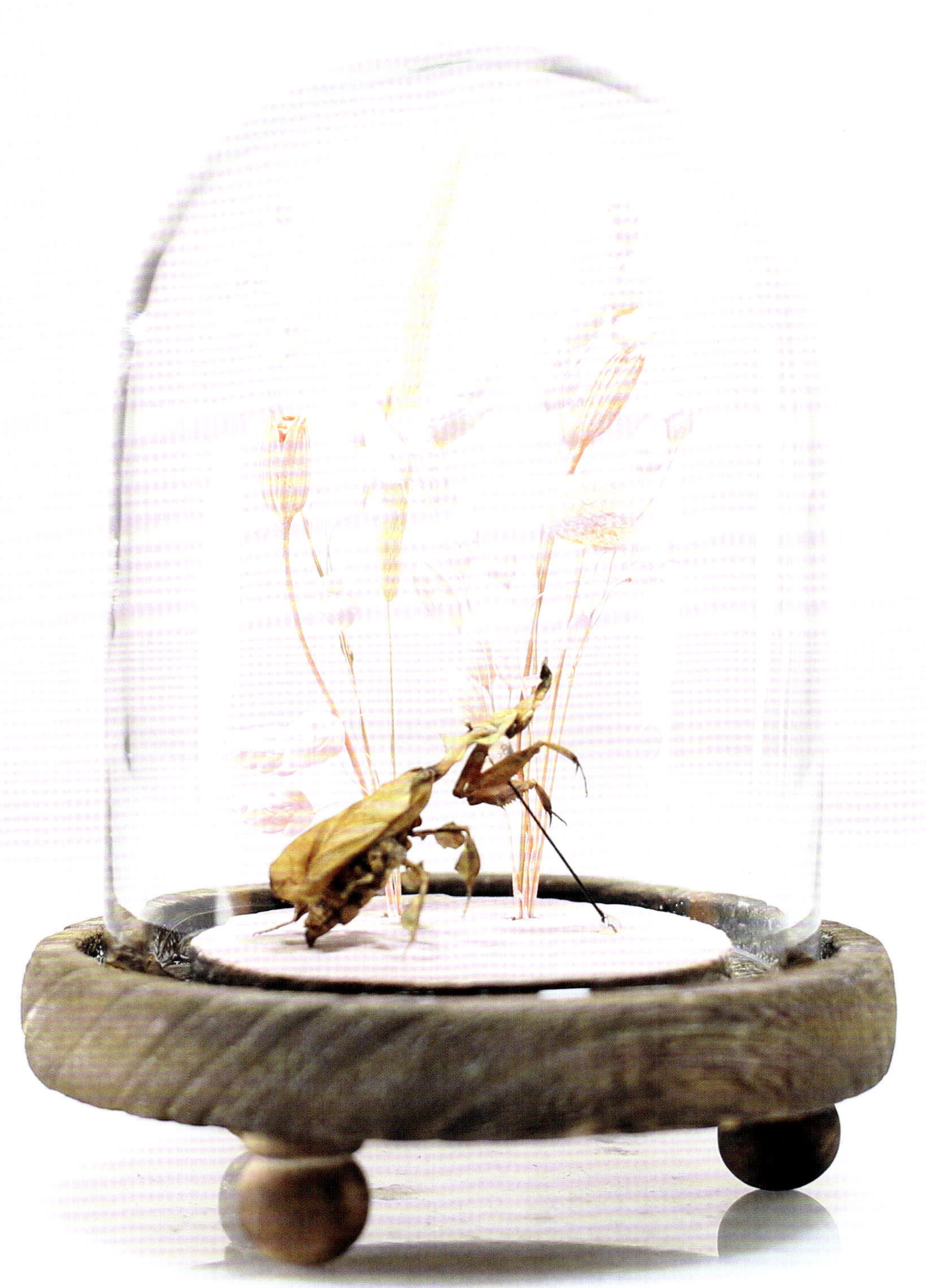

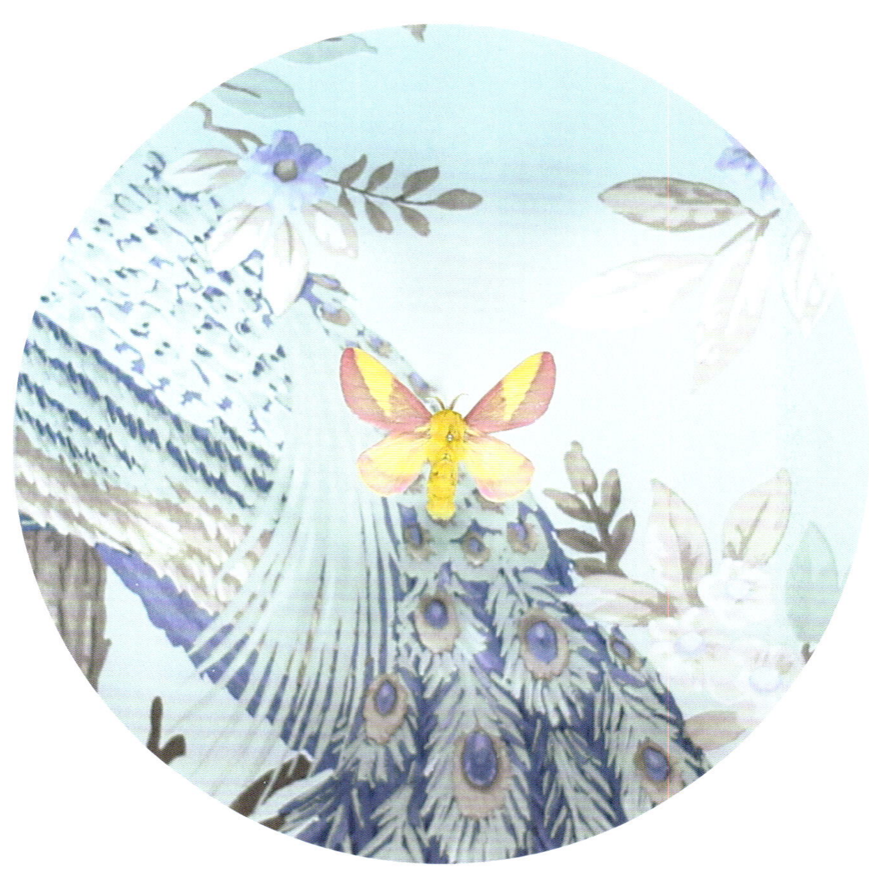

ROSY MAPLE MOTH

SCIENTIFIC NAME
Dryocampa rubicunda

GEOGRAPHIC RANGE
North America

This species is well known for its woolly body and gorgeous coloration, which can vary widely from soft cream or pure white to vibrant shades of bright pink and sunny yellow. These captivating hues most likely serve to warn off predators, and they may also help the adults blend in with the diverse colors of their flowering host trees.

THE ART of NATURE

Pattern in Insects

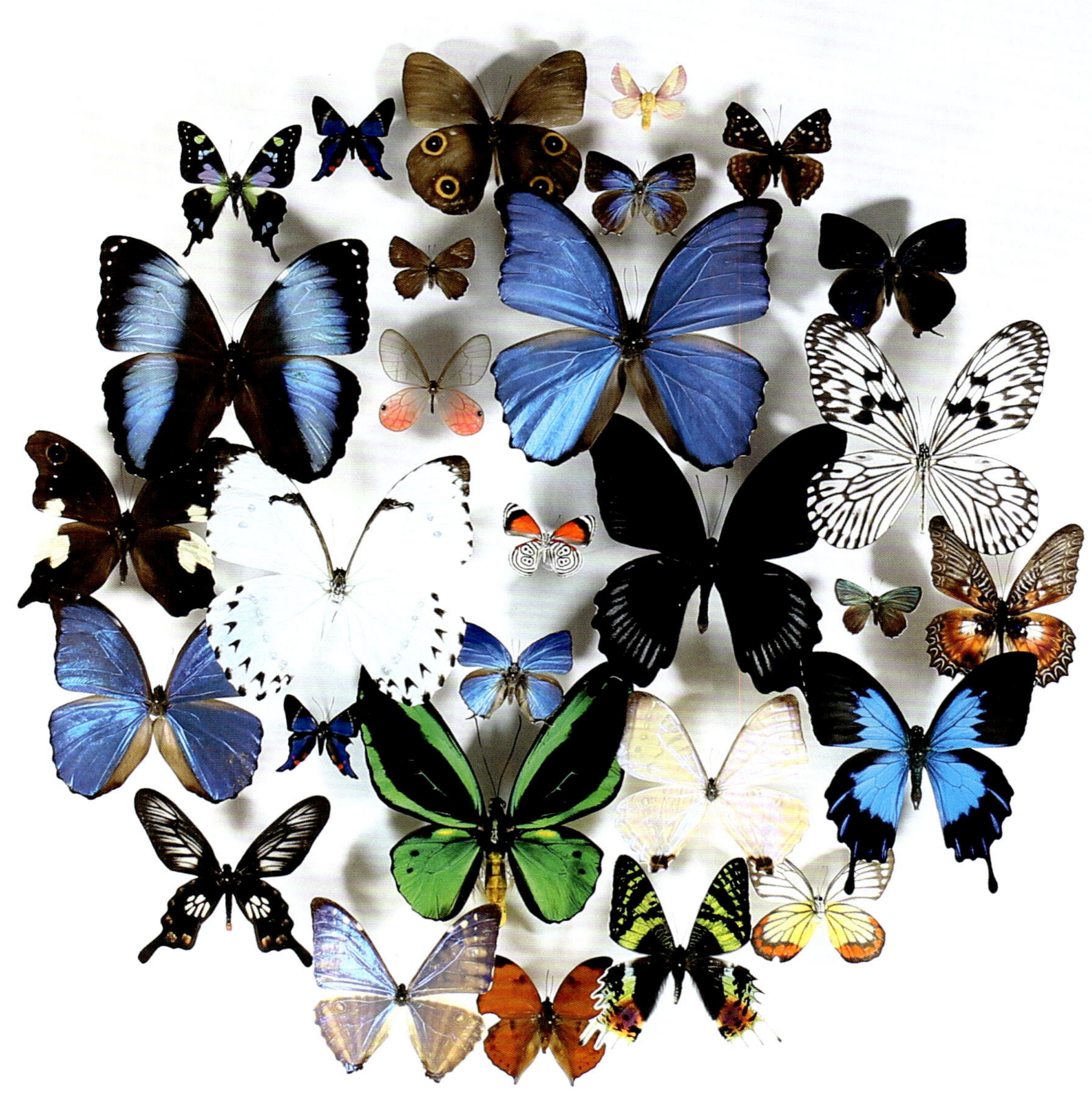

In the insect world, patterns are not merely ornamental; rather, they are often essential for survival. They manifest in an astounding array of forms: spots, stripes, bands, and even geometric shapes, each carrying a specific purpose. Camouflage is one of the most well-known strategies employed by insects, with many species adopting patterns that cleverly mimic the textures and colors of their specific environments—whether it be a leaf, the rough surface of bark, or the open expanse of the sky. The symmetrical beauty of certain butterflies' wings, for instance, helps them effectively disappear into the dappled light of the forest, protecting them from predators.

Sometimes, an insect's pattern serves the opposite purpose: It helps them stand out, signaling danger, asserting dominance, or attracting potential mates. The eye-catching black-and-yellow stripes of a wasp, for example, present an unmistakable warning: Approach at your own peril, as the consequences can be quite painful. In the realm of courtship, dazzling patterns like those displayed by the peacock butterfly or certain vibrantly colored moths send strong messages to potential mates, helping them recognize each other or even announcing their genetic fitness.

The complexity of these patterns, crafted through evolution over millions of years, suggests that nature is not only a master of survival but also a skilled artist, creating an extraordinary tapestry of life.

RAJAH BROOKE'S BIRDWING

SCIENTIFIC NAME
Trogonoptera brookiana

GEOGRAPHIC RANGE
Malaysia and Indonesia

Rajah Brooke's birdwing, a stunning representative of Malaysia's diverse wildlife, holds the title of the national butterfly of the country. The distinctive, pointed green shapes adorning the wings of this beautiful insect may help camouflage it by mimicking sharp thorns or the shapes of fern leaves.

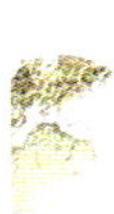

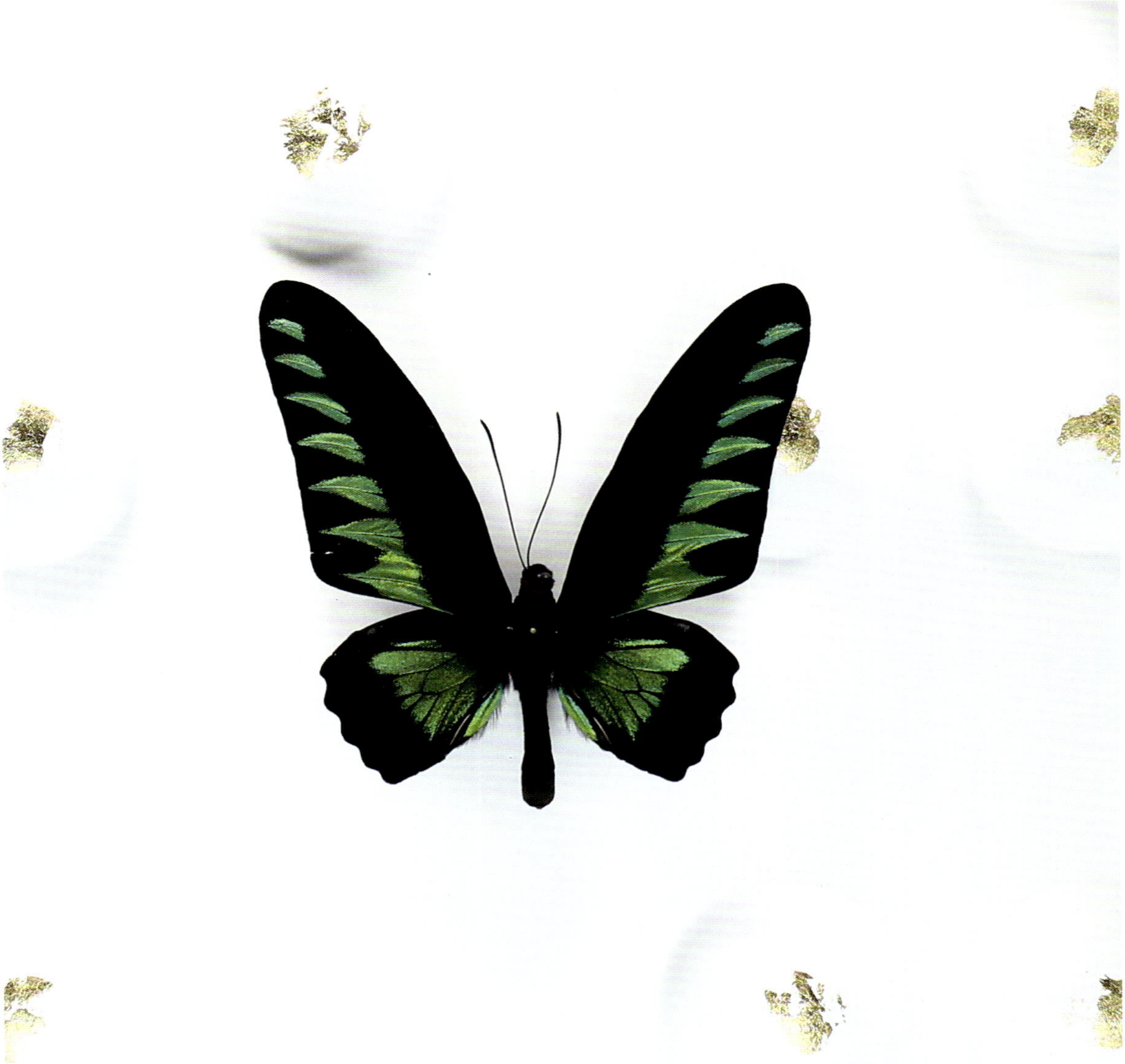

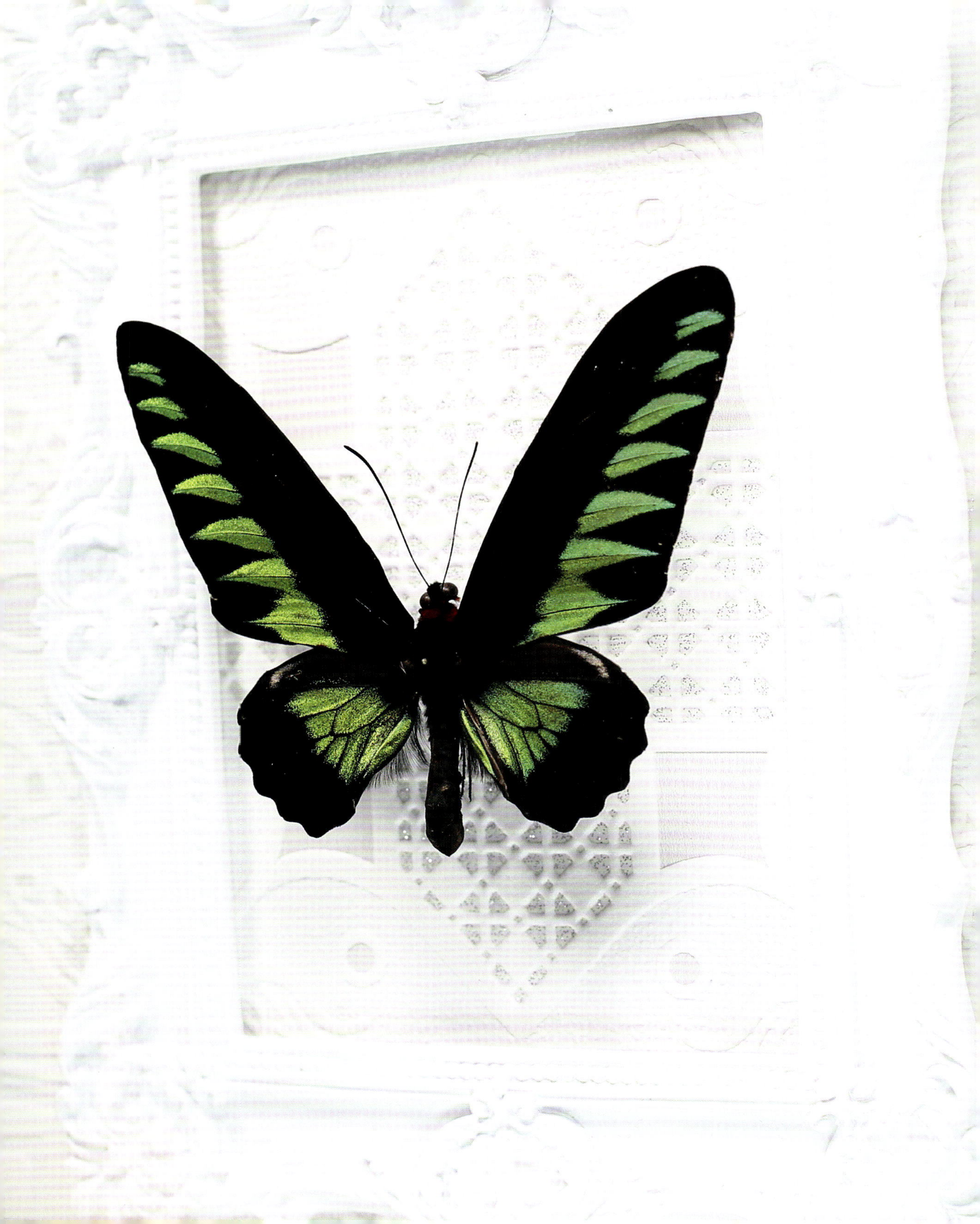

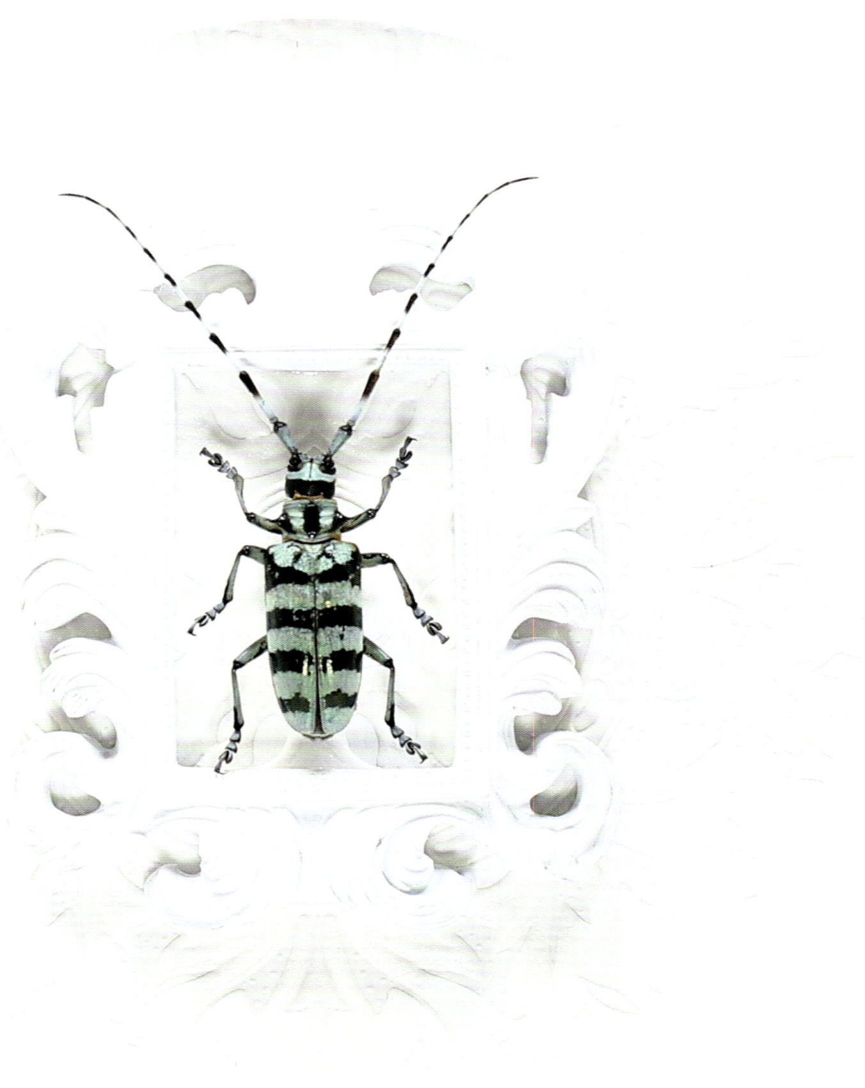

ZONATOR LONGHORN BEETLE

SCIENTIFIC NAME
Anoplophora zonator

GEOGRAPHIC RANGE
Southeast Asia

These large and colorful beetles are highly prized by collectors. Many members of this family of beetles are known to be pests, and their wood-boring larvae can cause significant damage and kill thousands of trees.

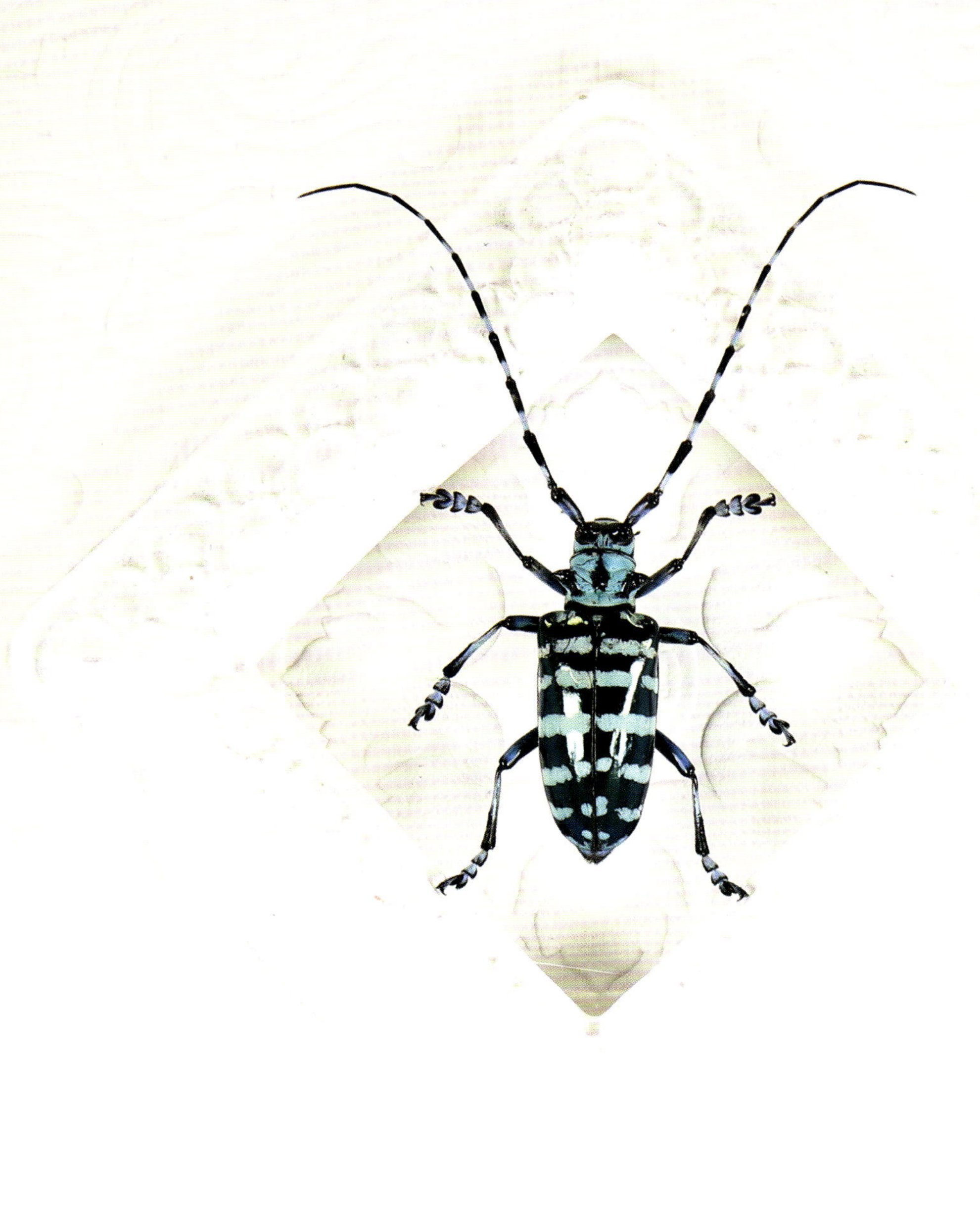

YELLOW UMBRELLA STICK INSECT

Scientific Name
Eurynecroscia nigrofasciata

Geographic Range
Malaysia and Indonesia

The yellow umbrella stick insect gets its name from its beautiful yellow-and-black wings, which closely resemble a paper umbrella when fully opened and displayed in all their glory. Females of this species are considerably larger than males, making the male quite a difficult gem to spot in its natural habitat.

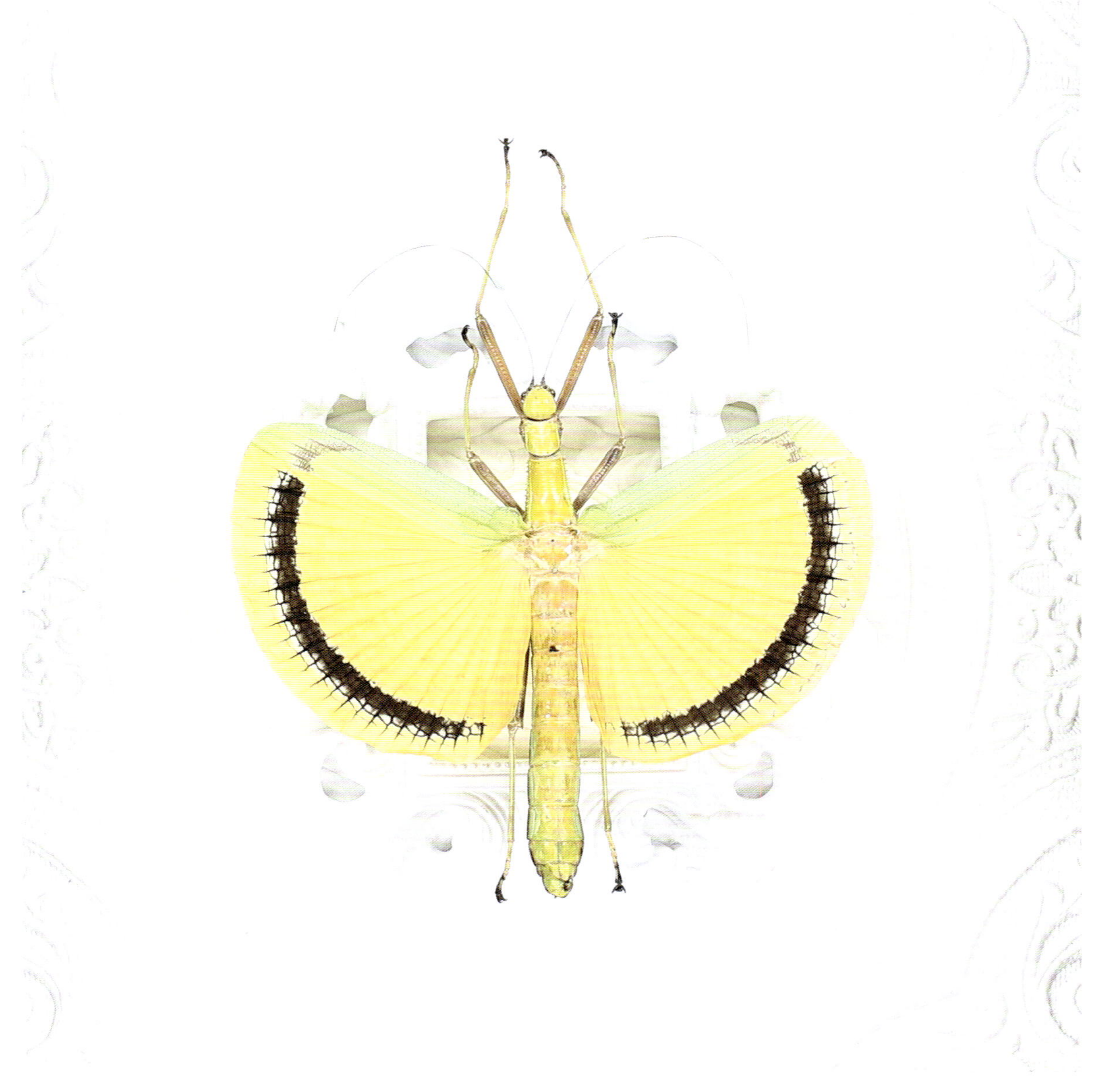

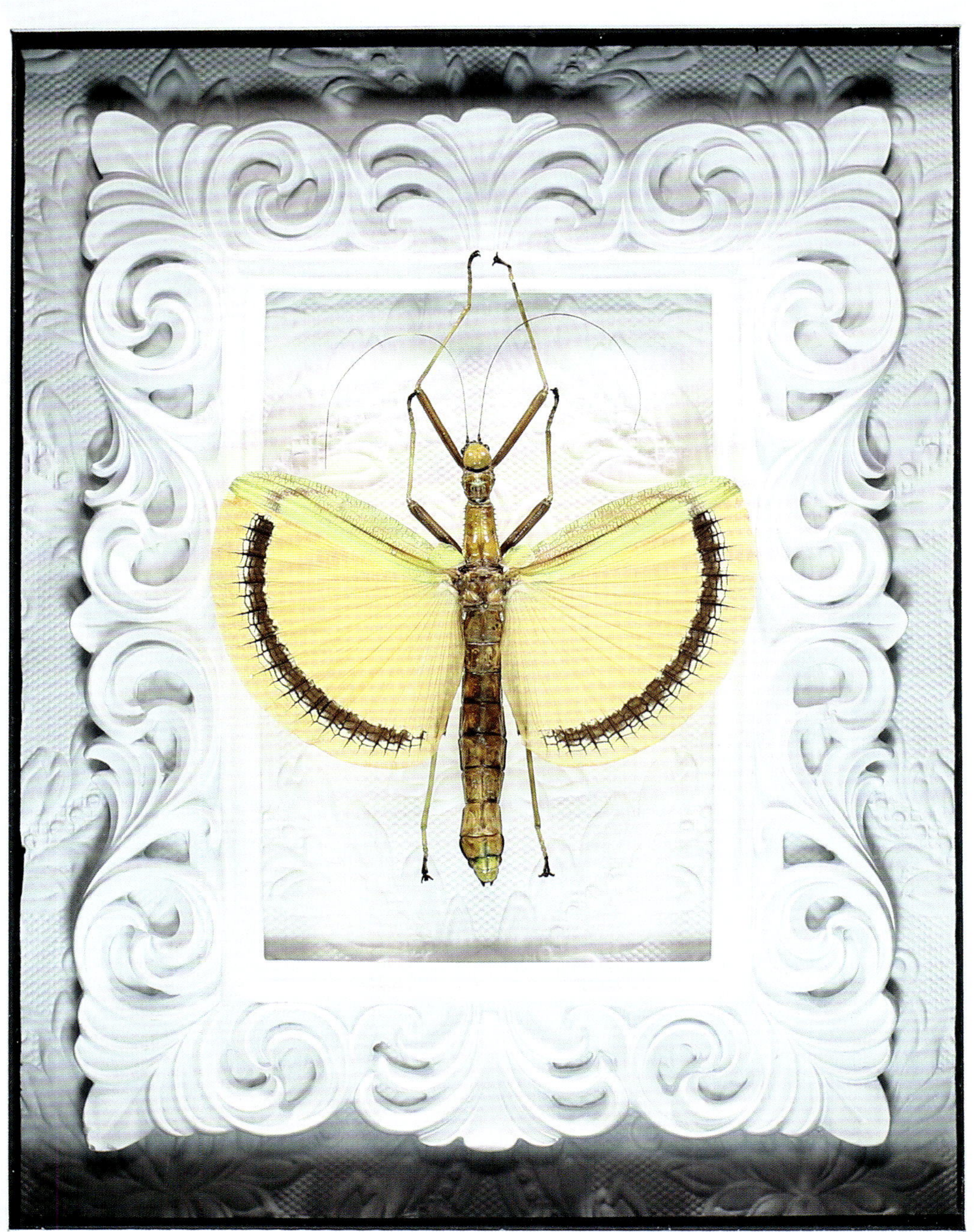

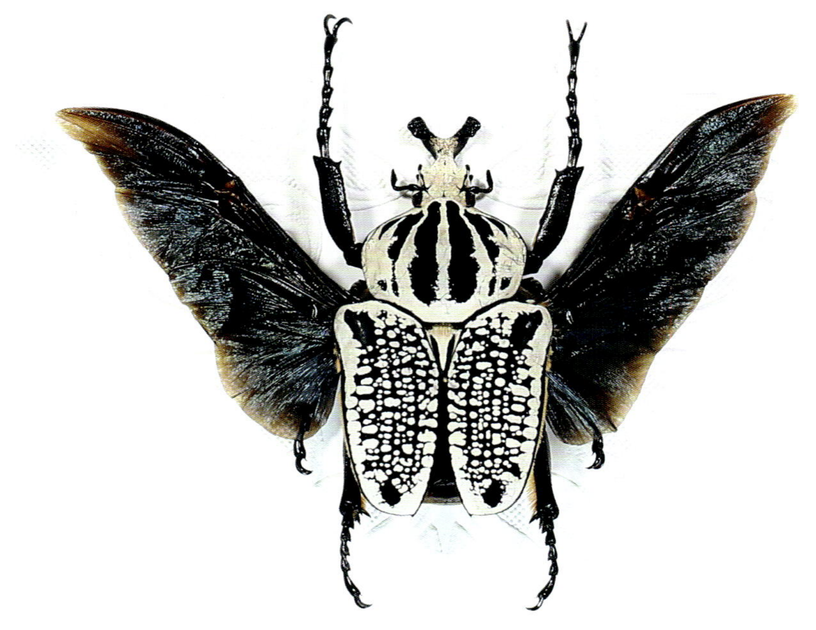

EASTERN GOLIATH BEETLE

SCIENTIFIC NAME
Goliathus orientalis

GEOGRAPHIC RANGE
Africa

The white areas on this beetle's body are more hydrophobic than the black parts. The pattern on this species might serve adaptive functions, such as the channeling of water off the body and thermoregulation.

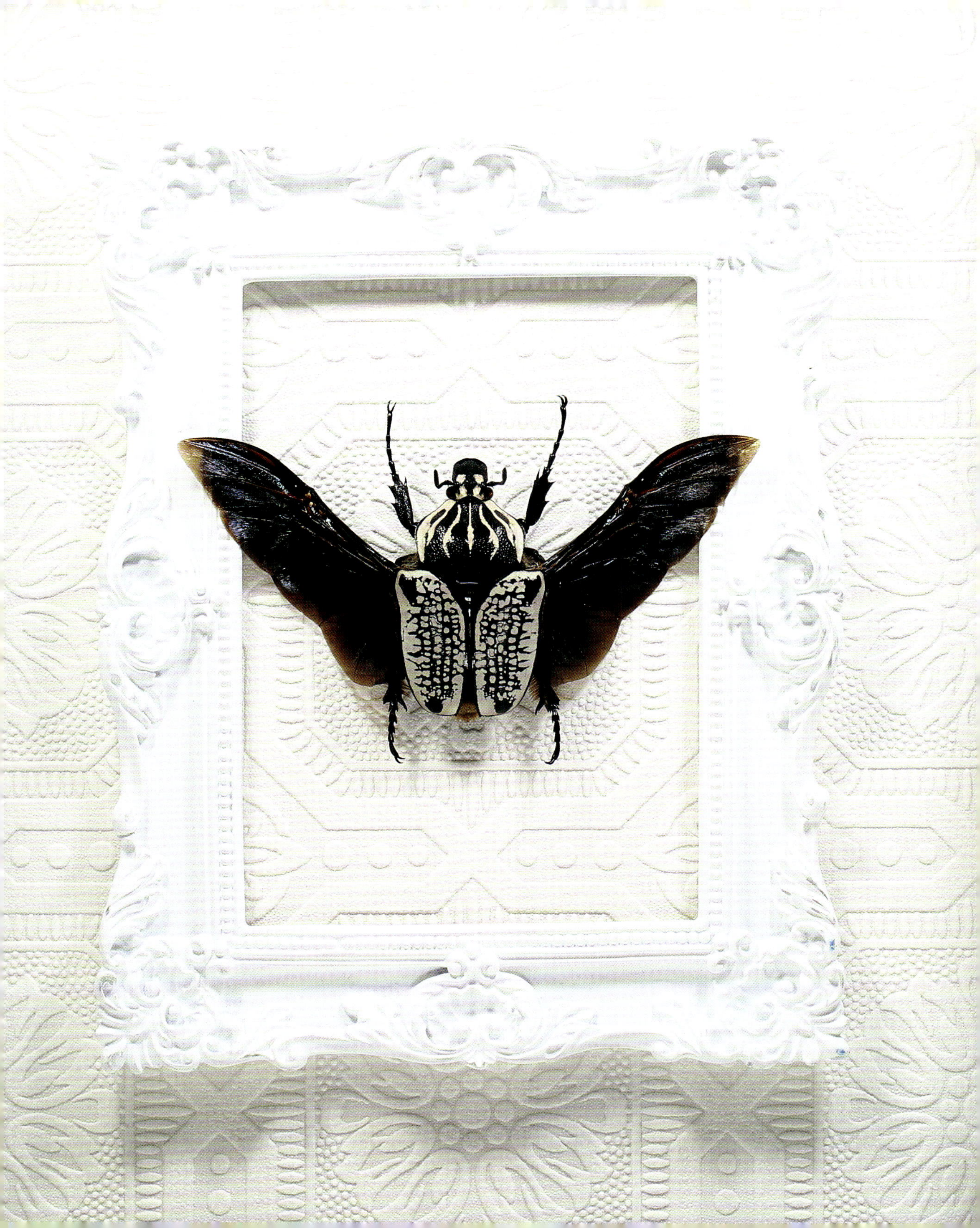

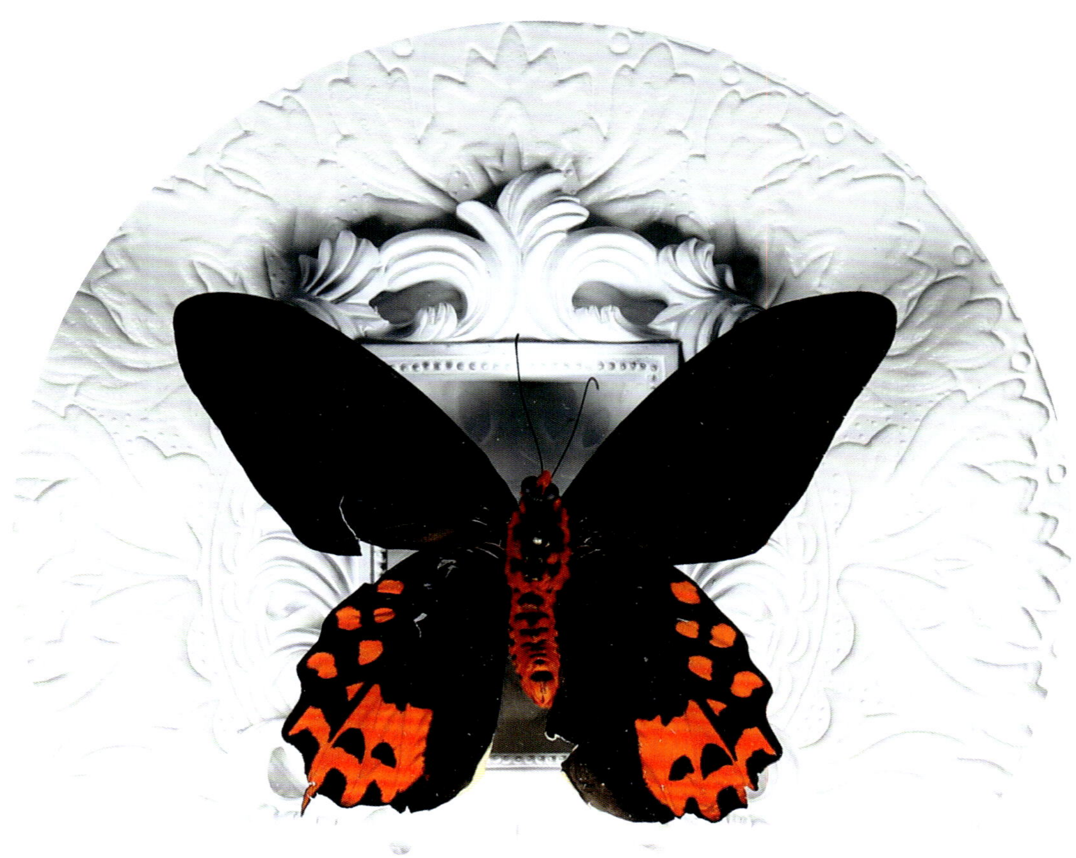

PHILIPPINE BATWING

SCIENTIFIC NAME
Atrophaneura semperi

GEOGRAPHIC RANGE
Philippines

Because they consume Indian birthwort while in the larval stage, adults in this species are distasteful to potential predators. Their bright red coloring is an effective warning that deters threats from would-be attackers.

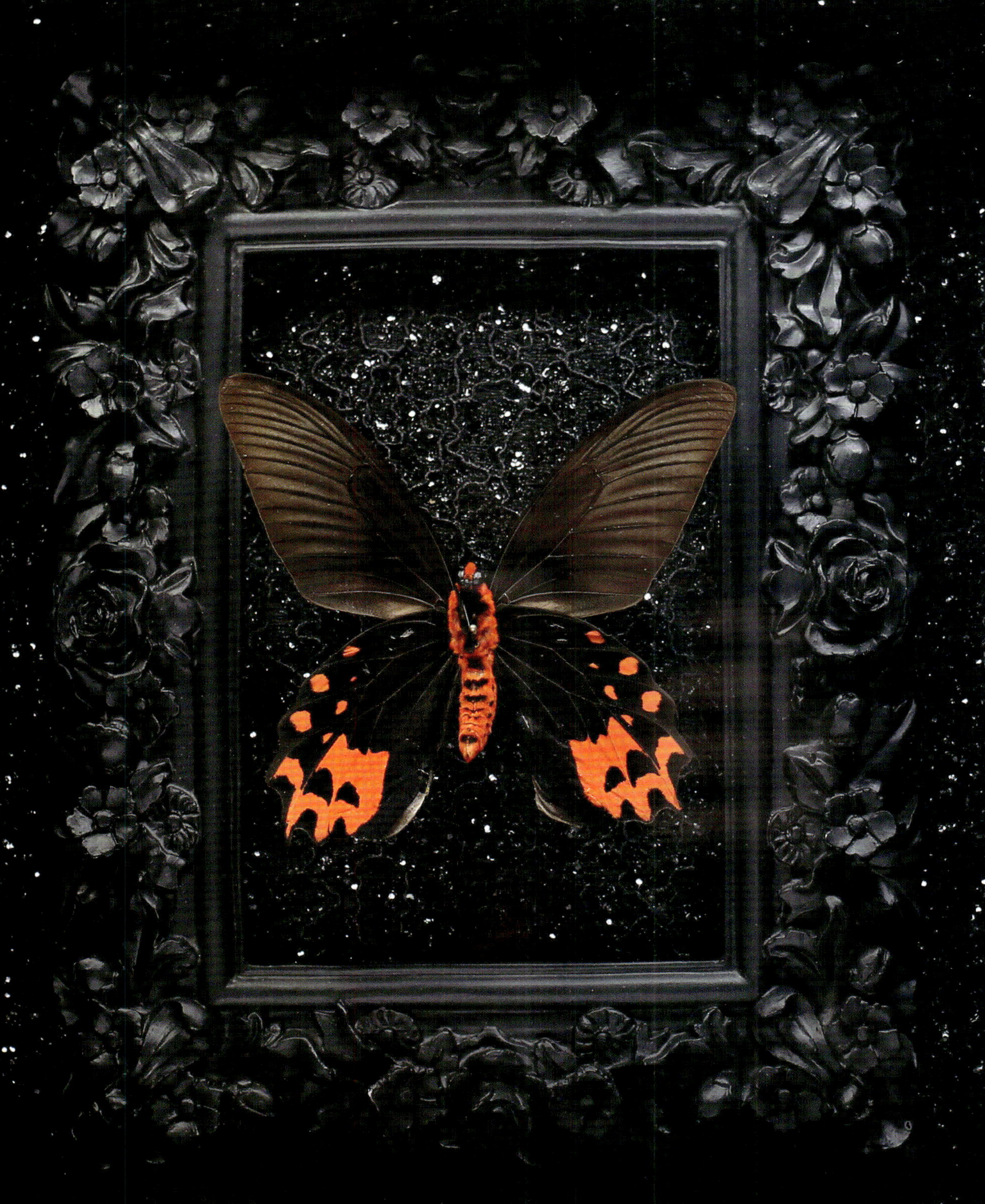

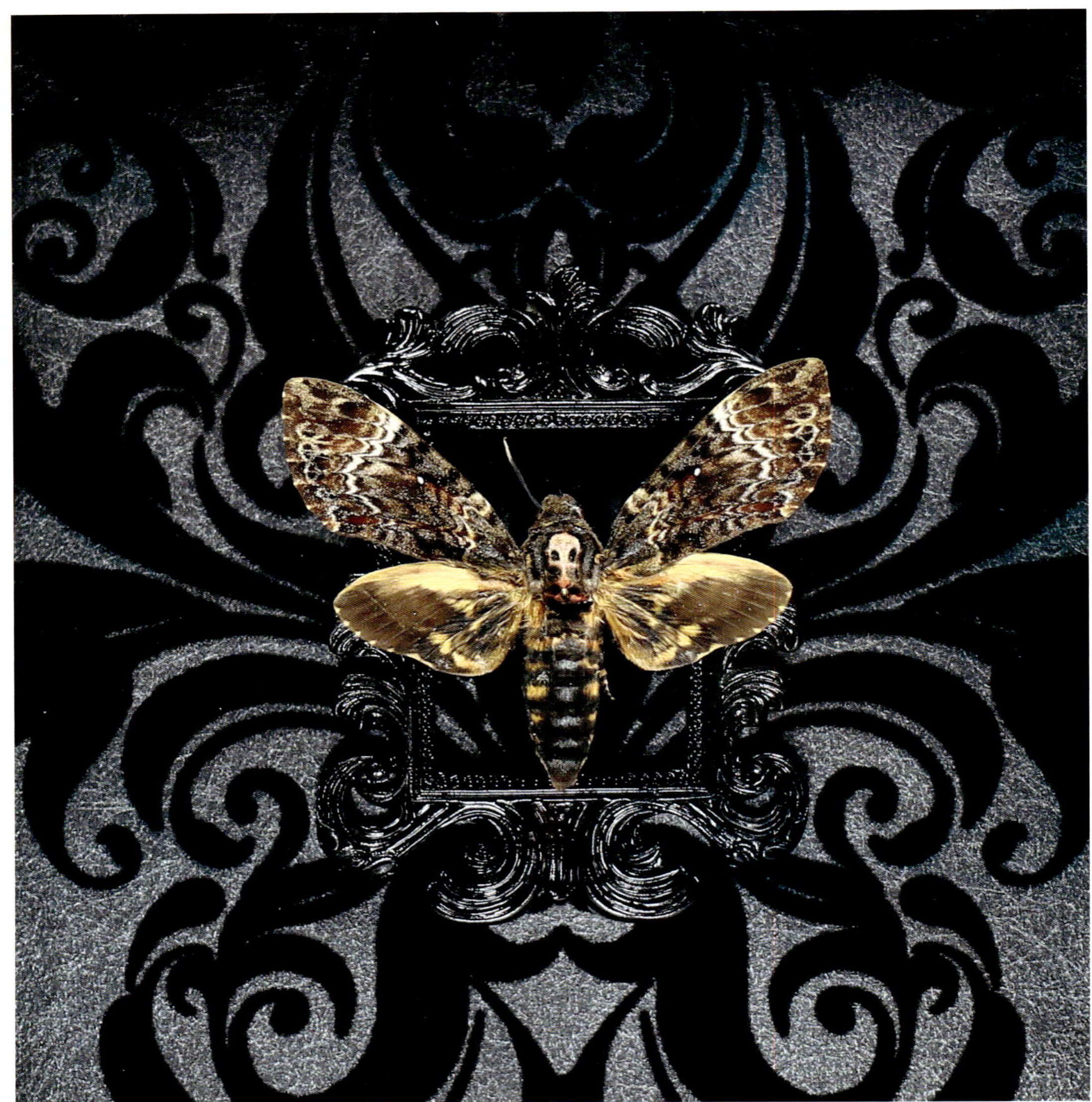

GREATER DEATH'S-HEAD HAWKMOTH

SCIENTIFIC NAME
Acherontia lachesis

GEOGRAPHIC RANGE
Asia

These moths have a distinctive human skull–shaped marking on the back of their heads. They have a notable fondness for honey, a vital part of their diet, often seeking it out during the night. Using chemical mimicry, they can even reproduce the scent of a bee colony, allowing them to enter hives and feed with minimal aggression from the resident bees.

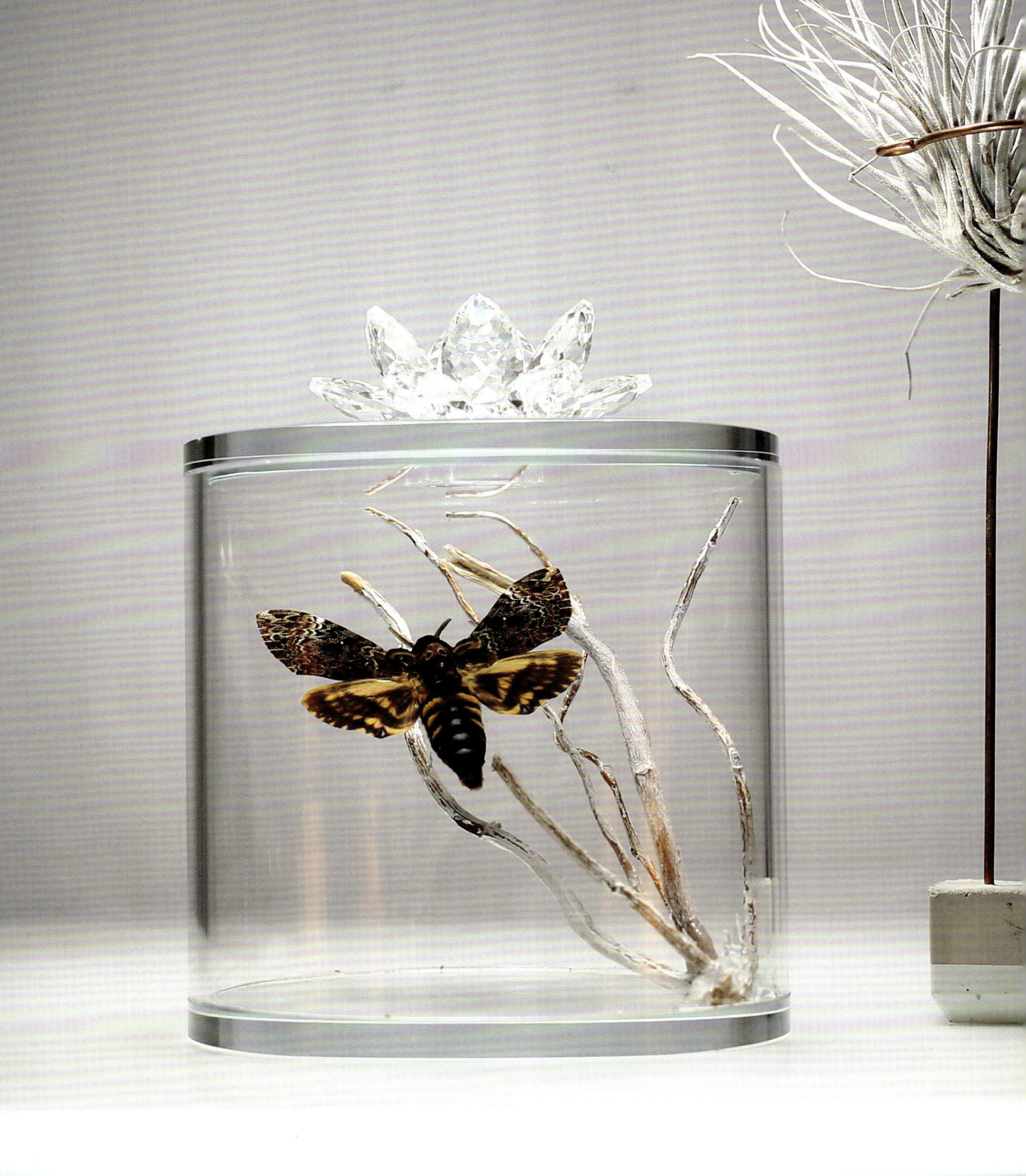

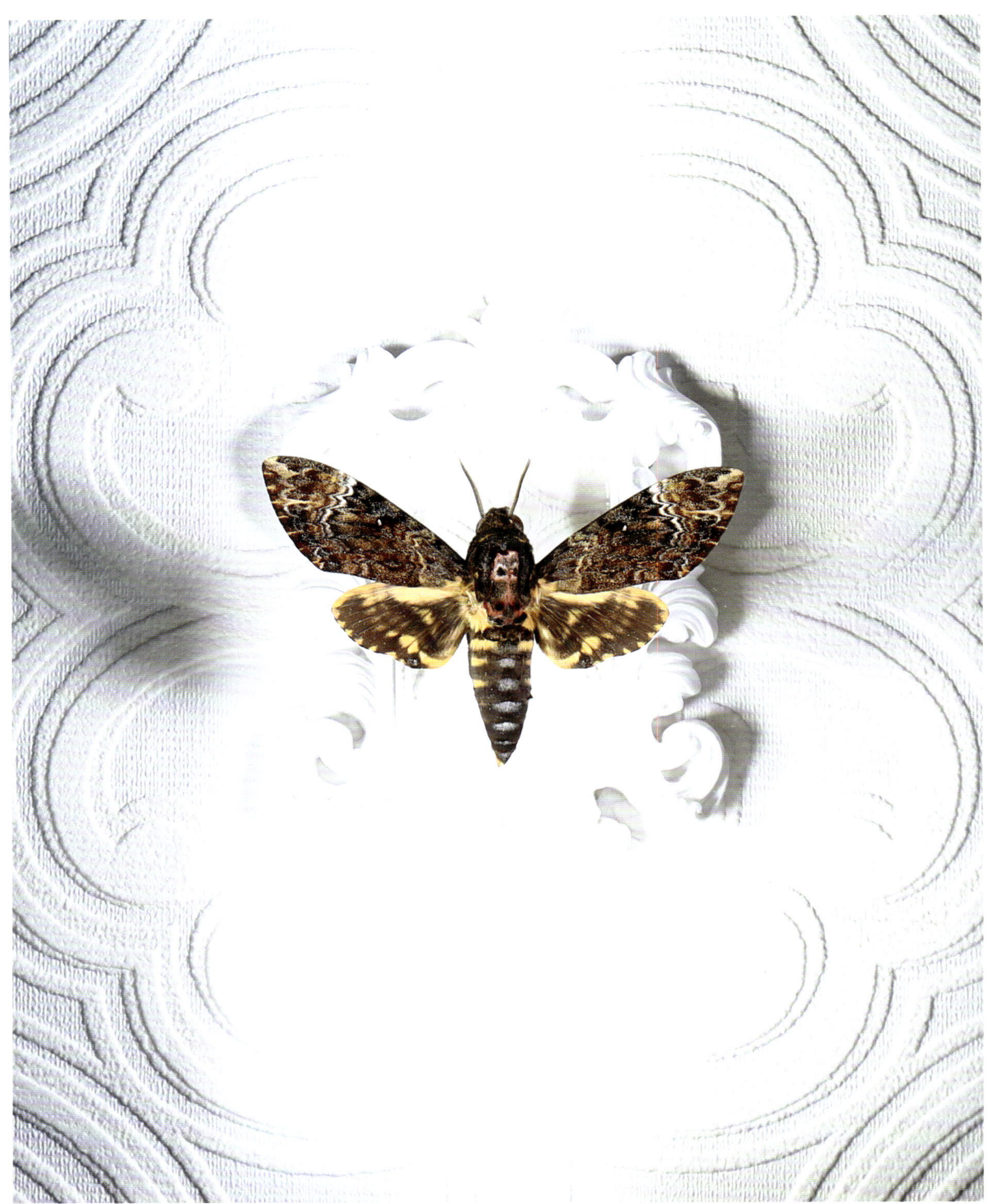

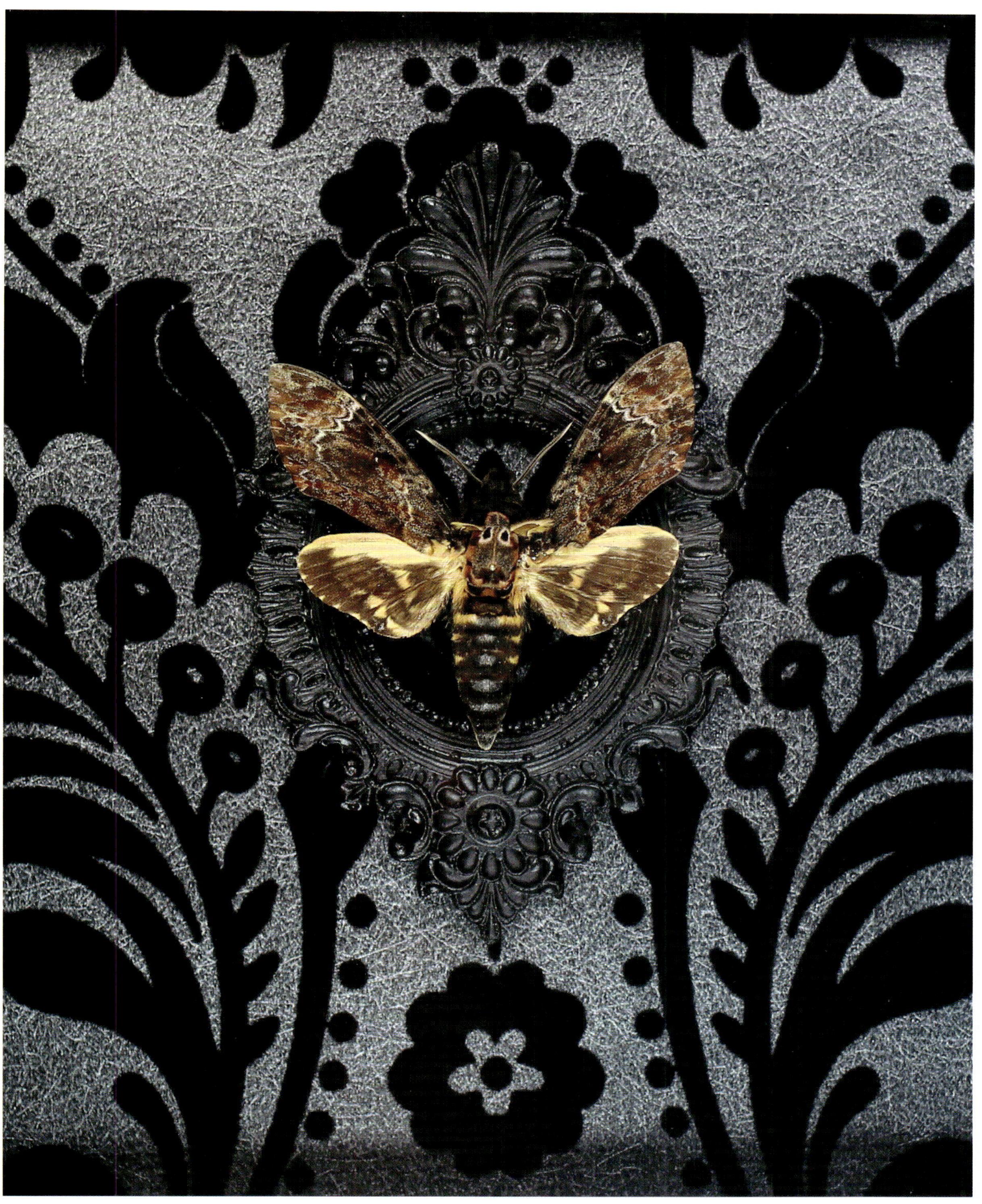

IO MOTH

SCIENTIFIC NAME
Automeris io

GEOGRAPHIC RANGE
North America and
Central America

The name of this moth comes from ancient Greek mythology, in which Io is depicted as a mortal who becomes the lover of the powerful god Zeus. Io moth caterpillars are covered with tiny, stinging spines, which can cause significant discomfort to humans and other mammals that come into contact with them.

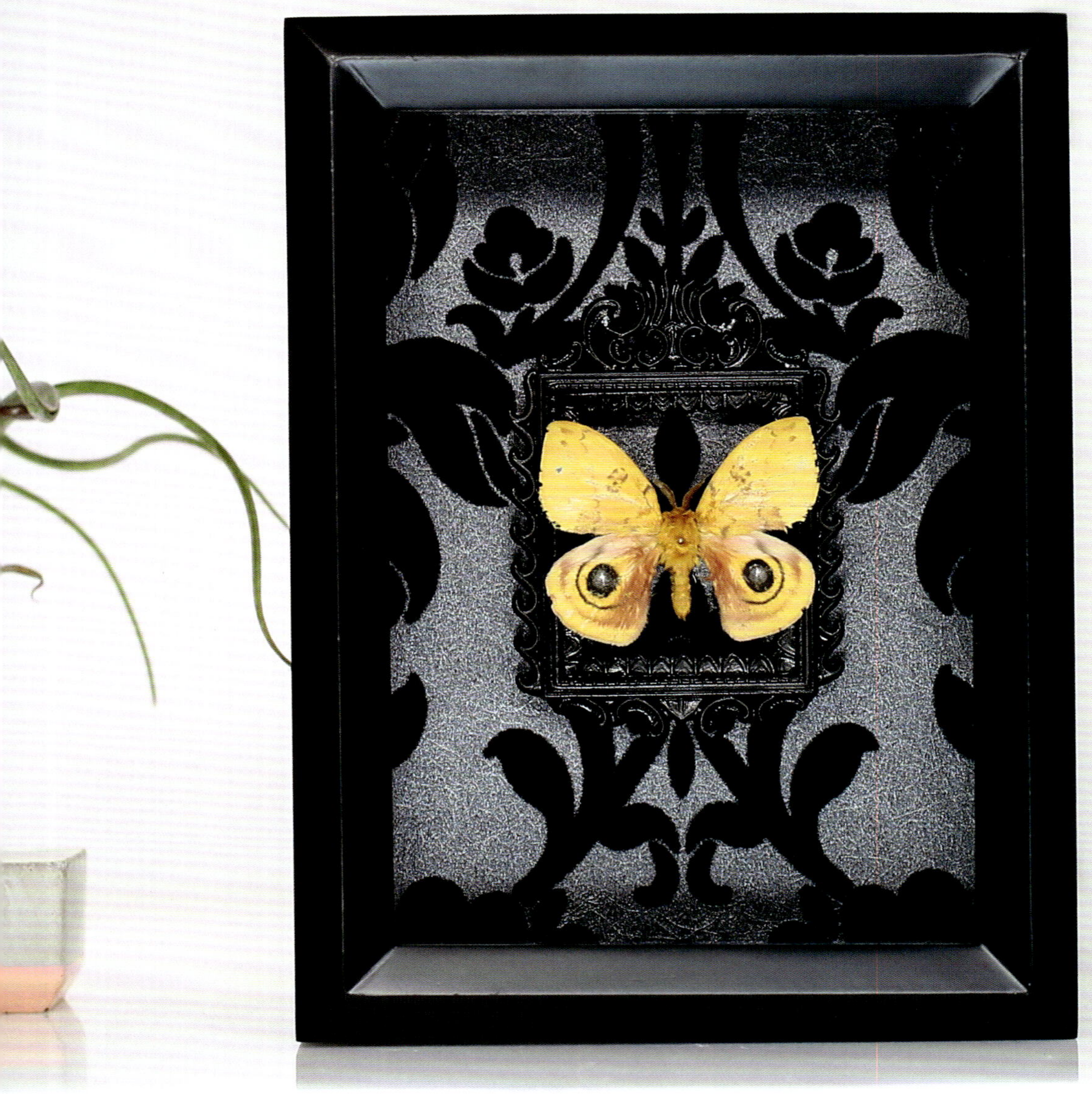

HARLEQUIN BEETLE

This beetle is very popular among collectors due to its vibrant, colorful, and unusual appearance. During the mating season, males use their remarkably long forelimbs to engage in competitive bouts, attempting to dislodge rivals from dead or decaying trees. These trees are later chosen by females as egg-laying sites, offering vital nutrients for their developing larvae.

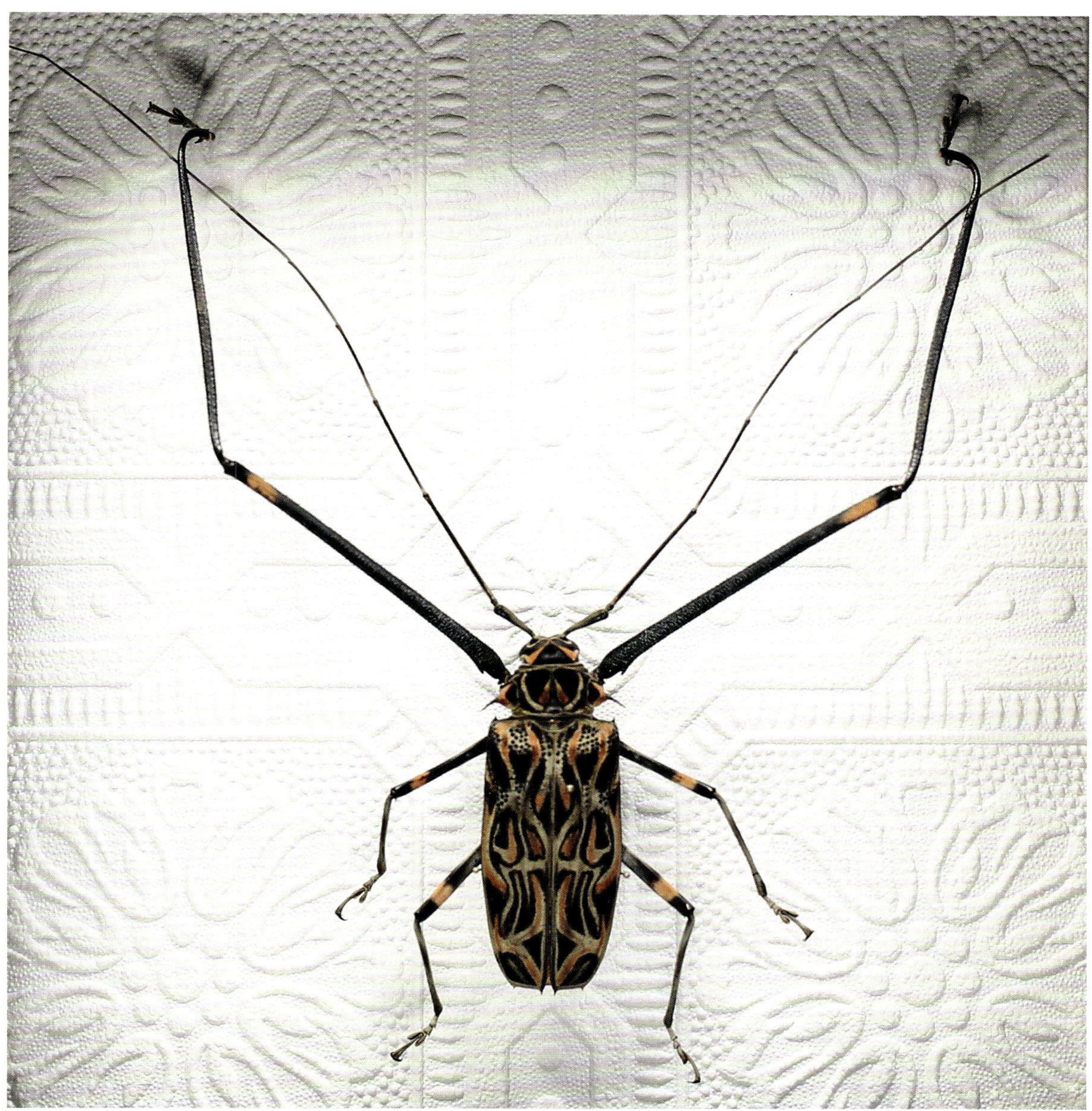

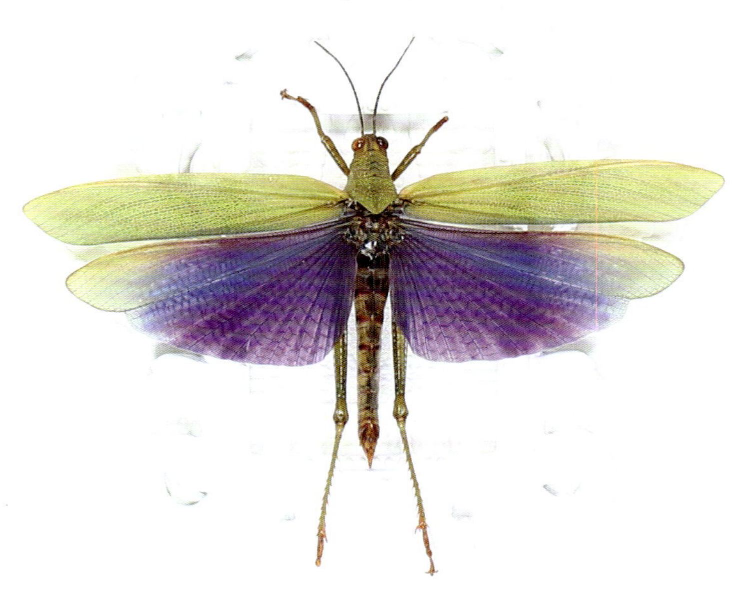

PURPLE-WINGED GRASSHOPPER

SCIENTIFIC NAME
Titanacris albipes

GEOGRAPHIC RANGE
South America,
especially Amazonia

At rest, the purple-winged grasshopper displays cryptic green forewings that blend into the foliage of the upper canopy. But when it takes flight, it exposes vivid violet or magenta hind wings—a startle mechanism that deters predators.

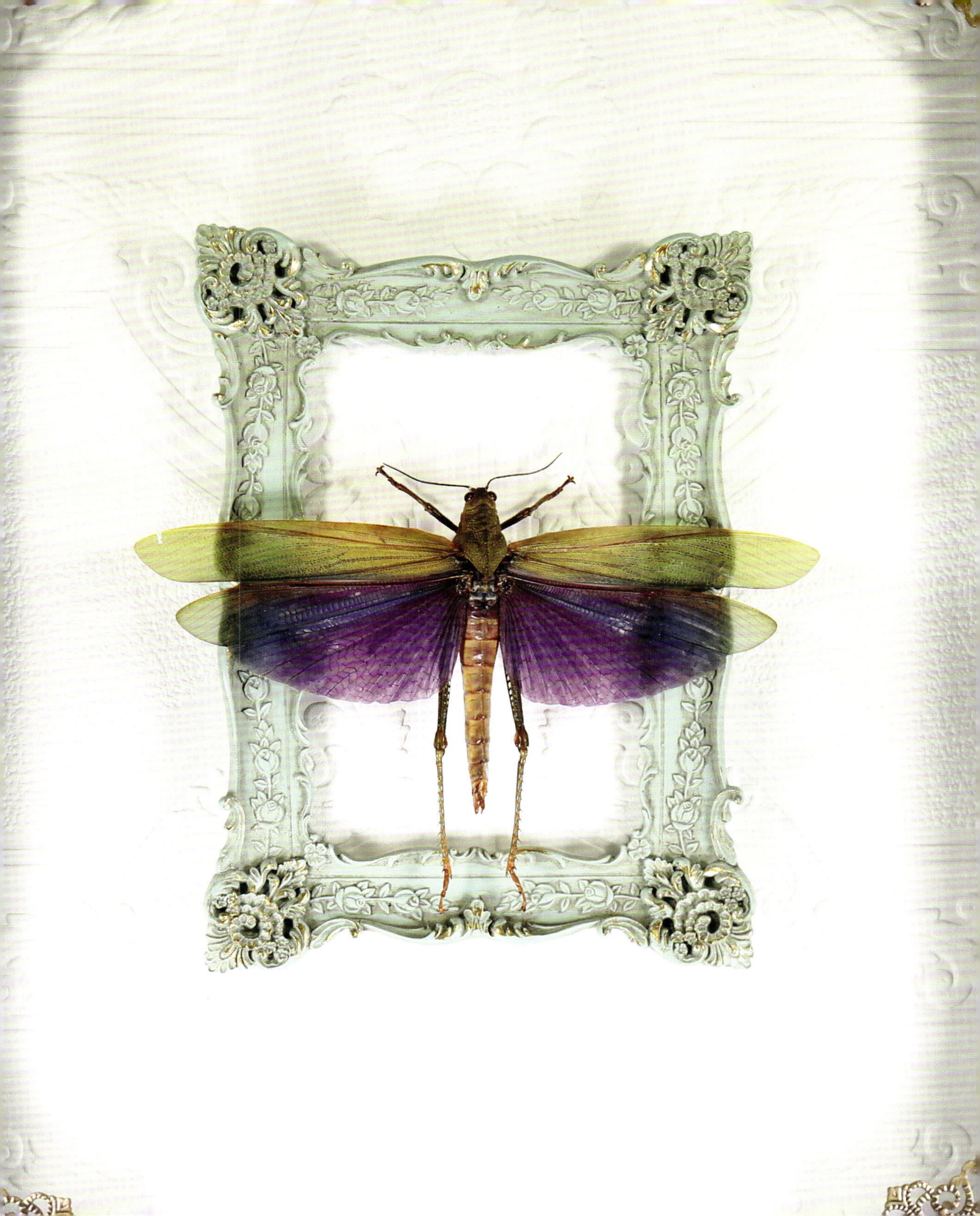

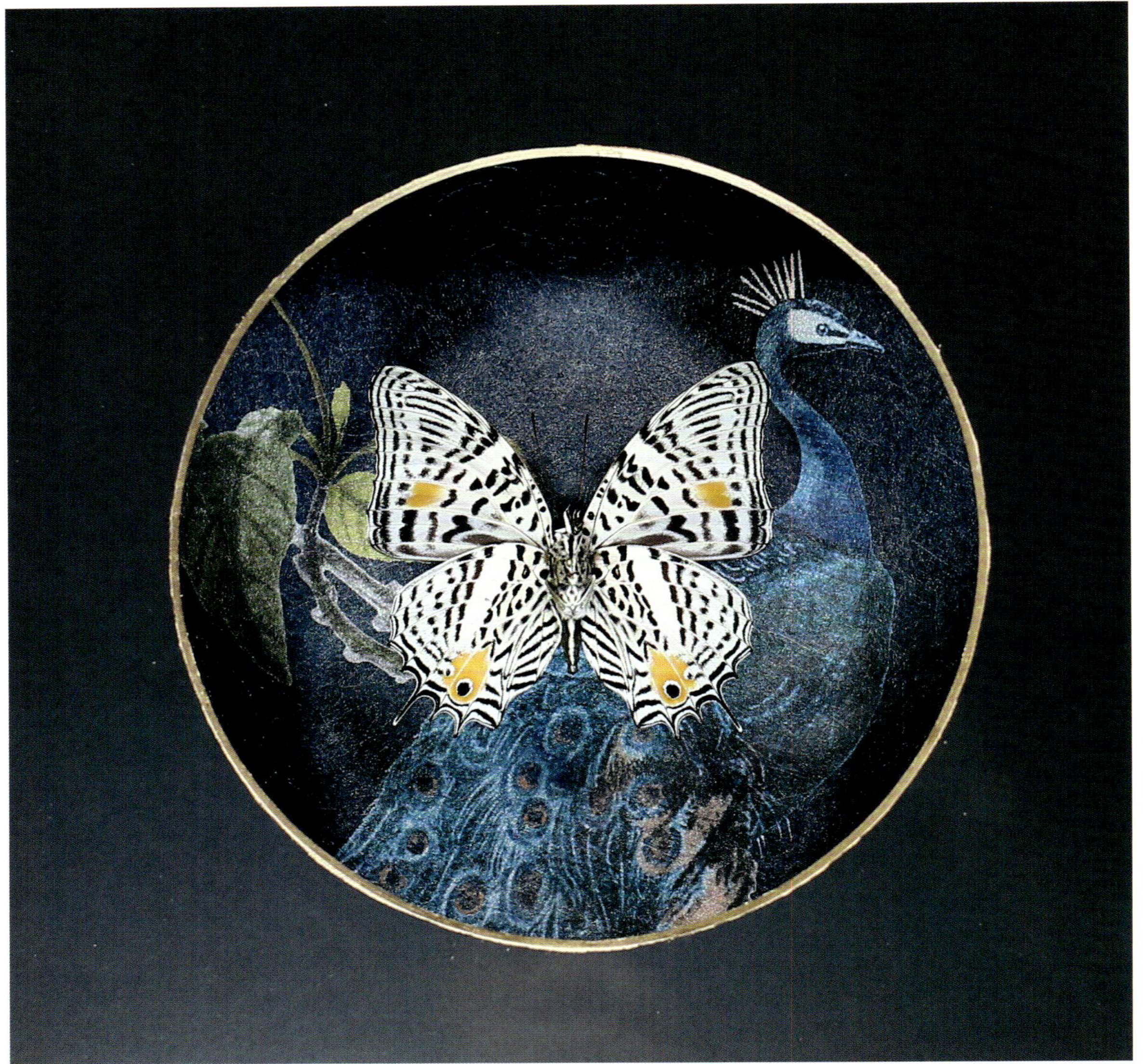

AMAZON BEAUTY

SCIENTIFIC NAME
Baeotus aeilus

GEOGRAPHIC RANGE
Ecuador, Peru, and Brazil

This butterfly exhibits a notable case of sexual dimorphism. The male's top is characterized by white and dark brown coloring complemented by four orange eyespots. The female is also quite striking, adorned with a banding of pale orange that enhances her overall beauty and makes her easily distinguishable.

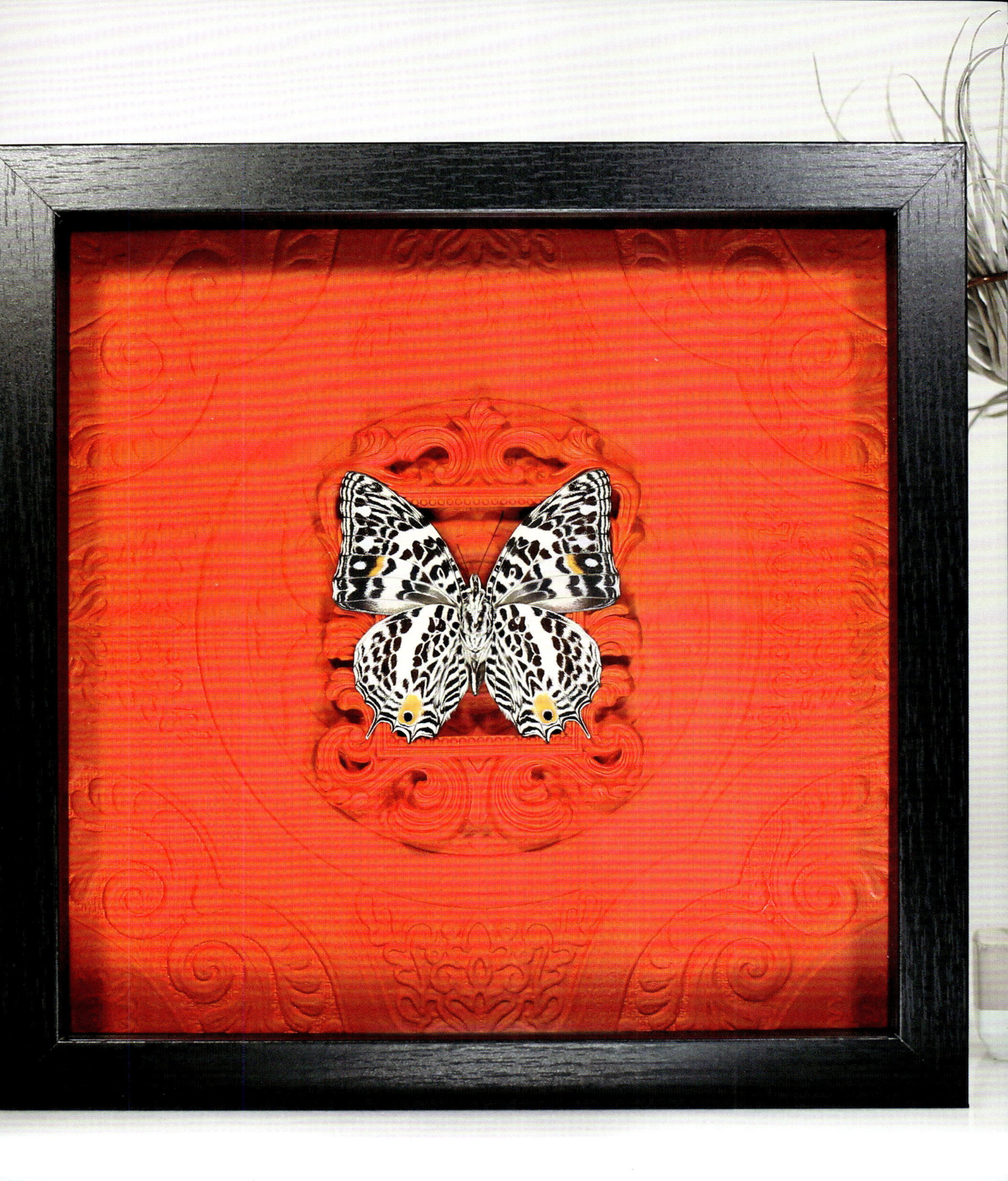

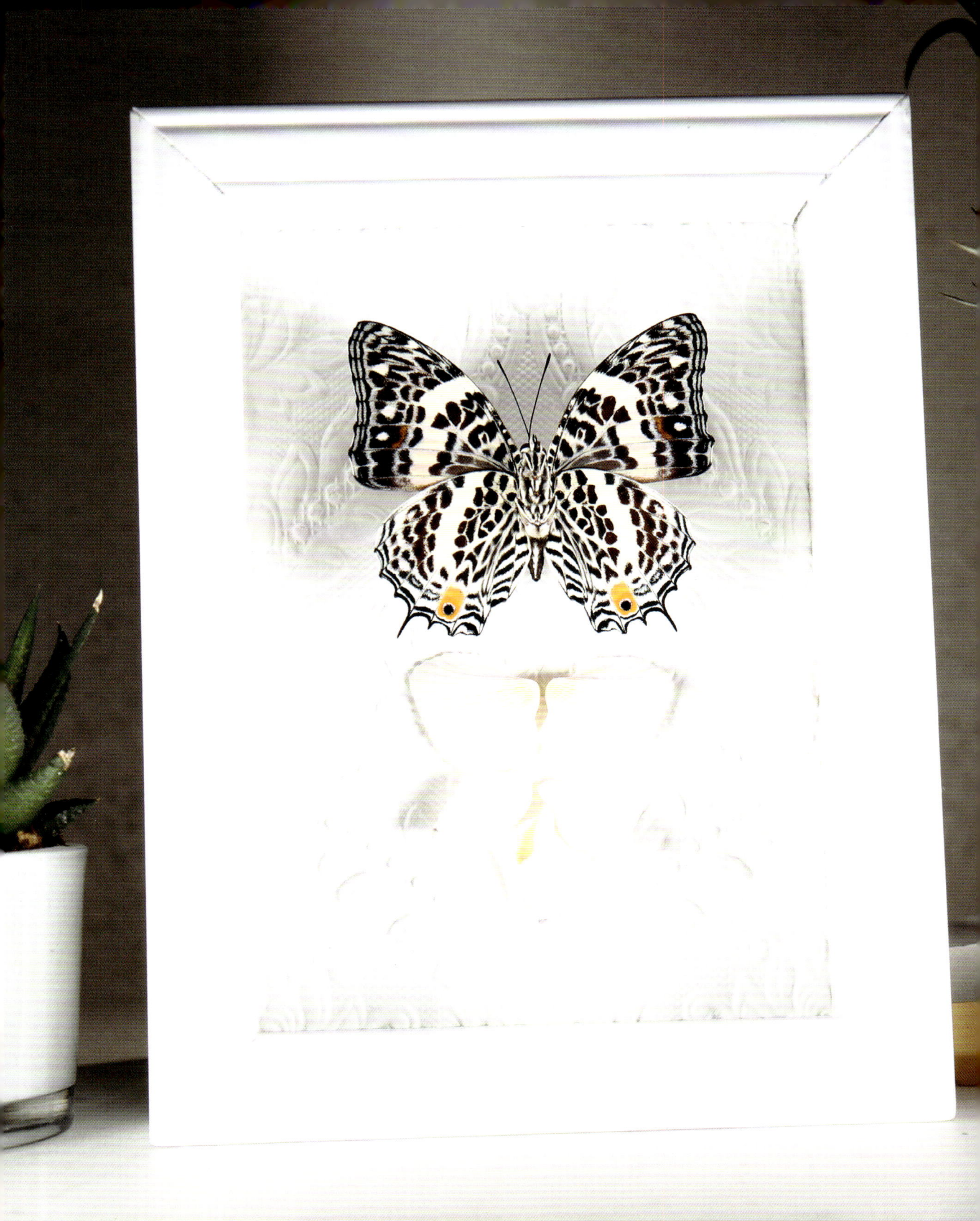

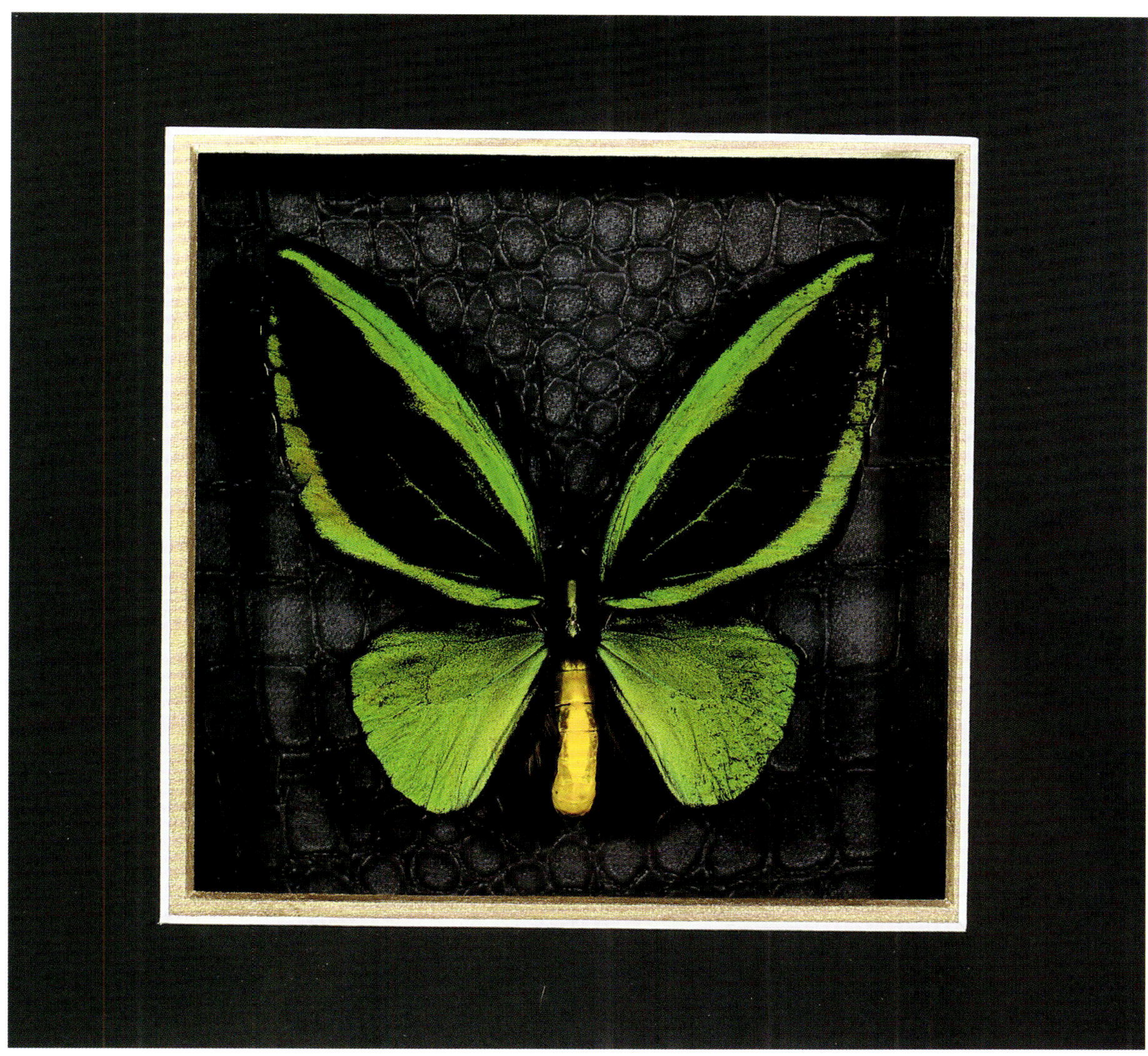

CAPE YORK BIRDWING

SCIENTIFIC NAME
Ornithoptera priamus

GEOGRAPHIC RANGE
Moluccas, New Guinea,
Bismarck Archipelago,
Solomon Islands, and
northeastern Australia

These butterflies belong to the Papilionidae family, which contains the largest butterflies in the world. The scientific binomen of this species, *Ornithoptera priamus*, comes from the name Priam, referring to the Trojan king of the ancient Greek epic *The Iliad*.

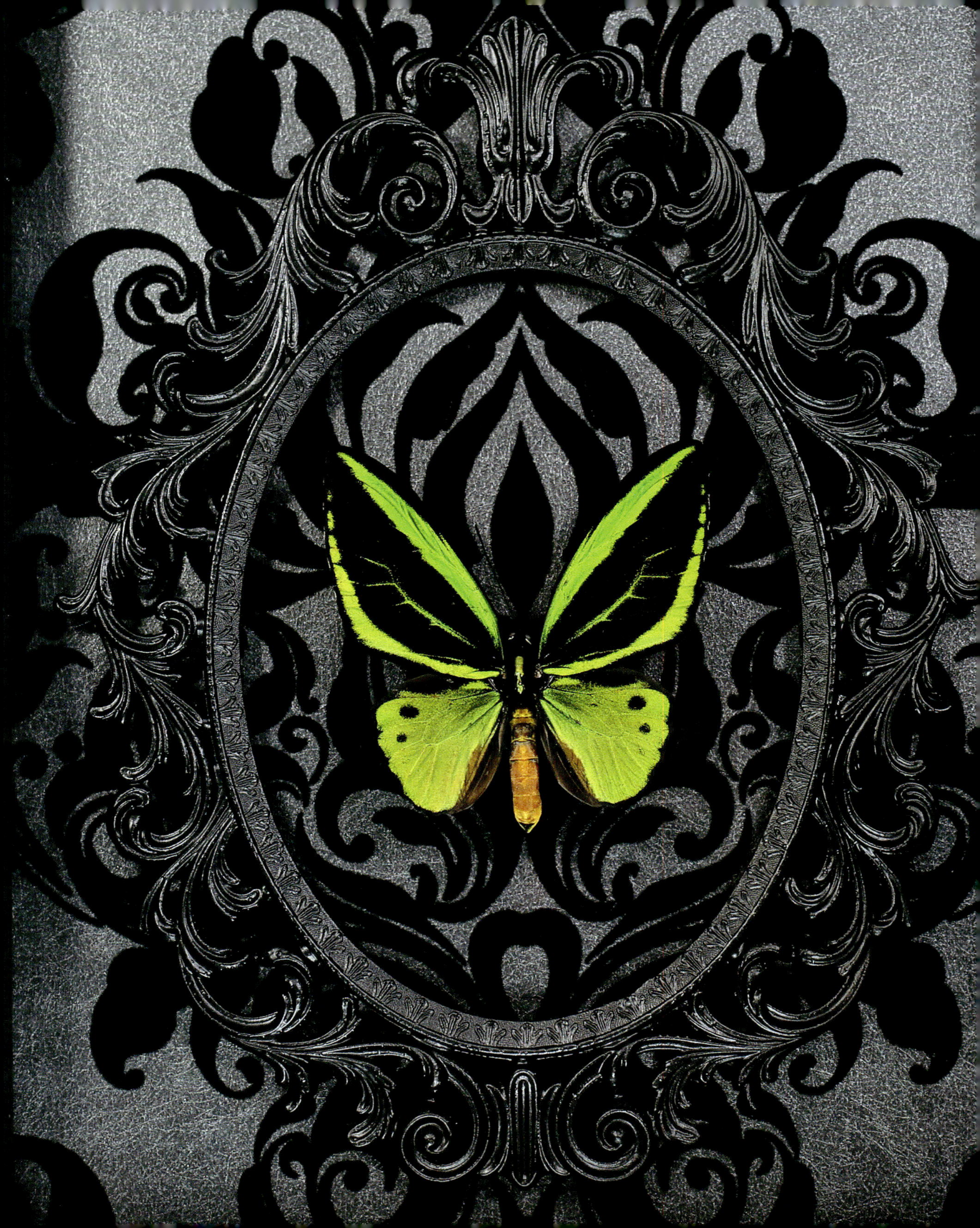

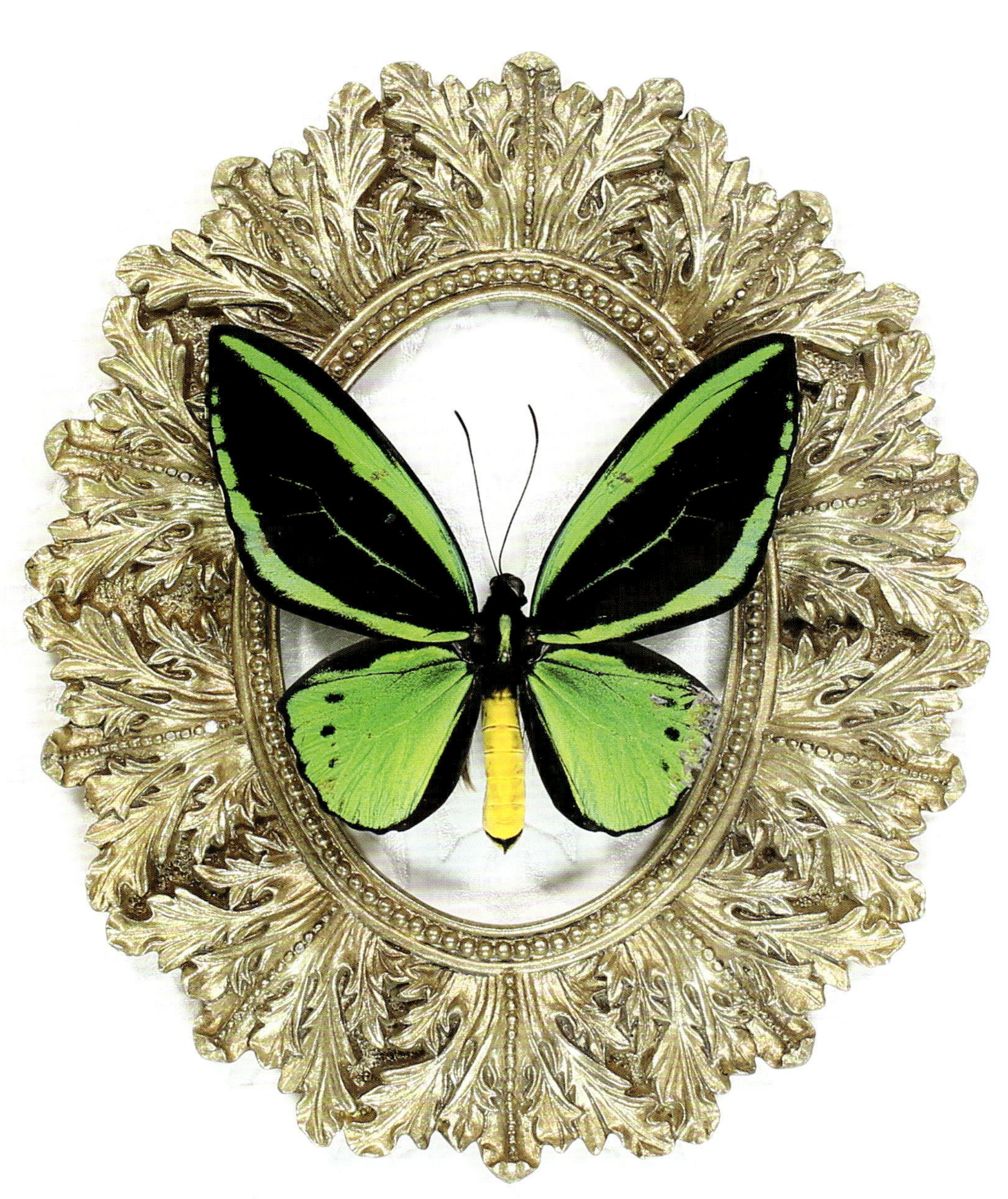

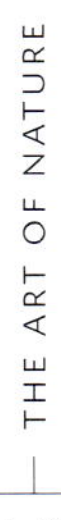

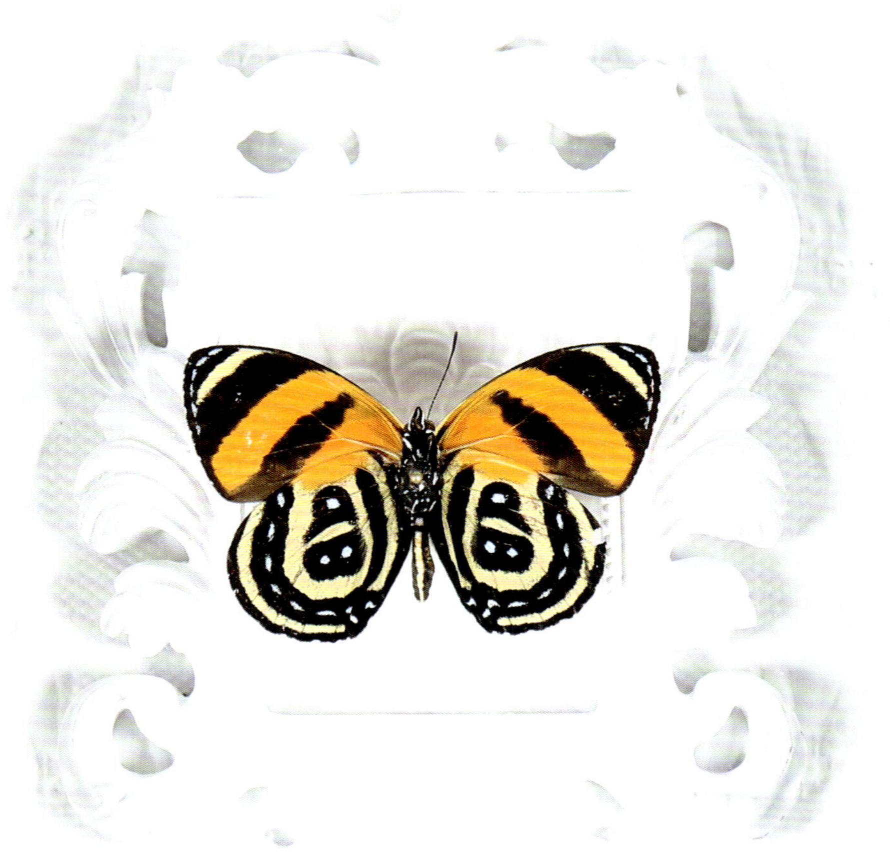

BD BUTTERFLY

SCIENTIFIC NAME
Callicore cynosura

GEOGRAPHIC RANGE
South America,
especially Amazonia

The common name of this species is derived from the bold markings found on the underside of members' hind wings, which resemble the letters *B* and *D*. In hot weather, they fly in search of salt and other minerals, which they often find on sand banks or on the sweaty skin of the humans they land on.

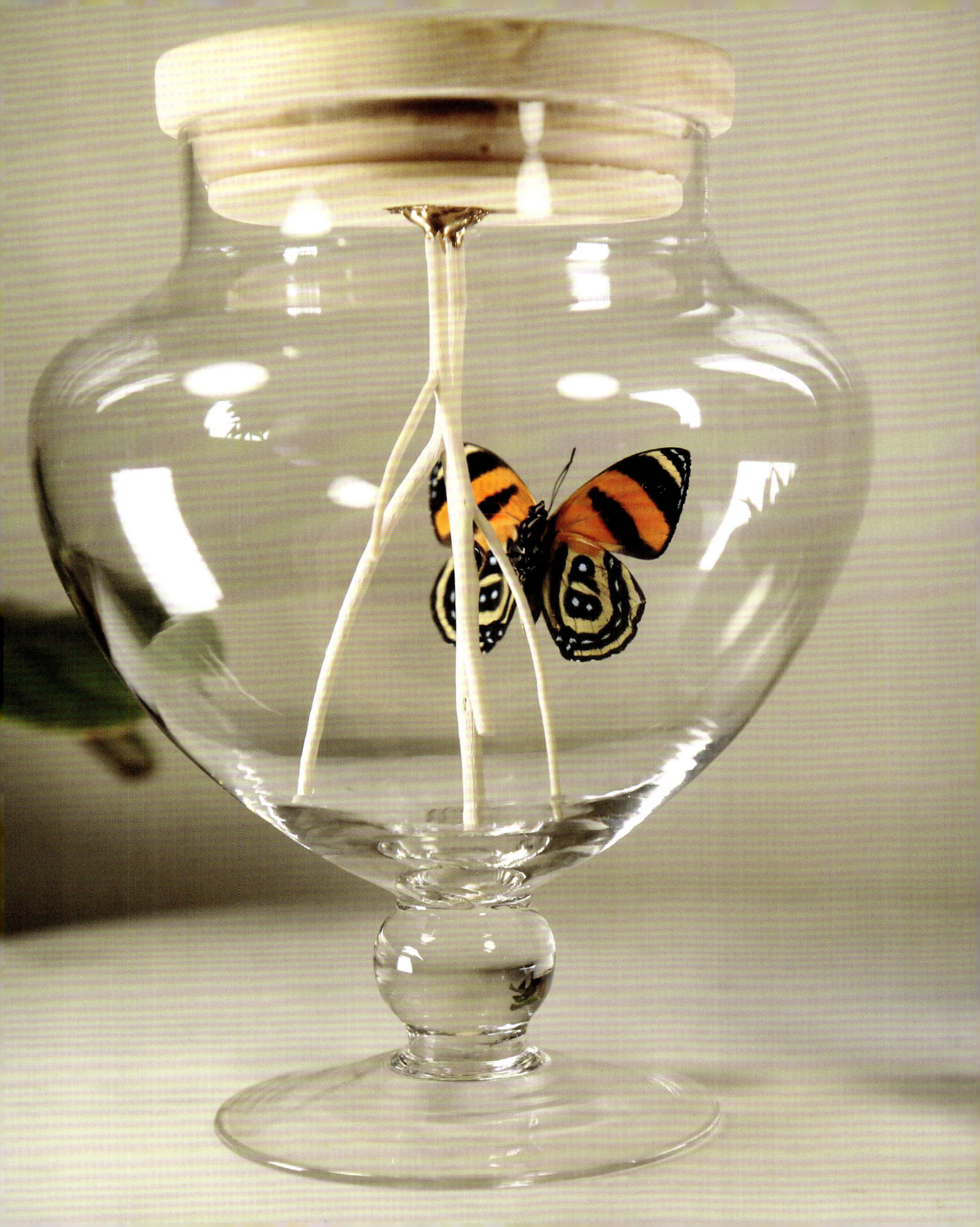

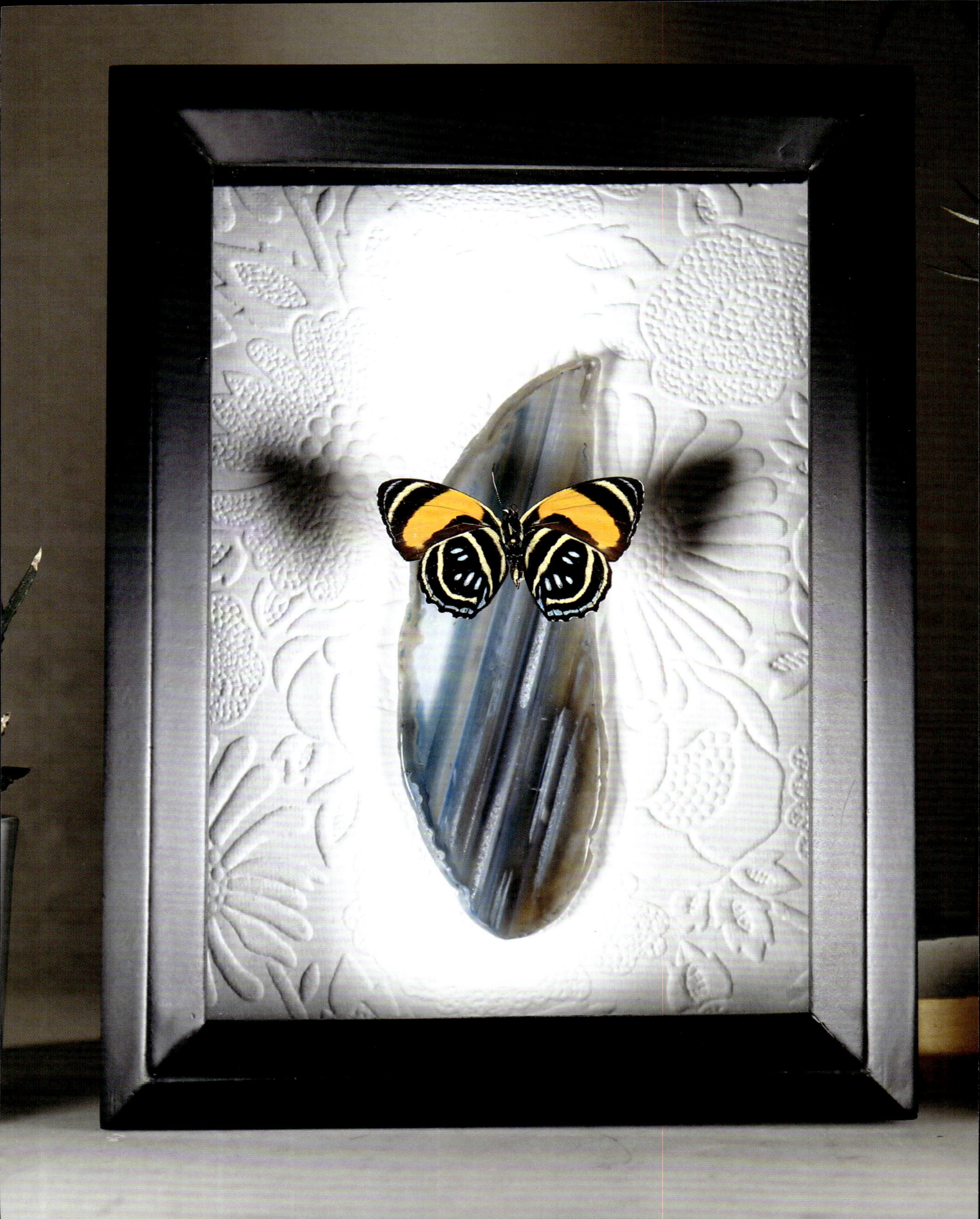

ORANGE WALKING STICK

Paracyphocrania major

GEOGRAPHIC RANGE
Indonesia

This noteworthy walking stick camouflages into the dense canopy of the forest. If its appearance as a branch or stick fails to deceive potential threats, it has an alternative defense mechanism. Like the purple-winged grasshopper, it attempts to scare off predators by flashing its bright orange wings, creating a startling visual display.

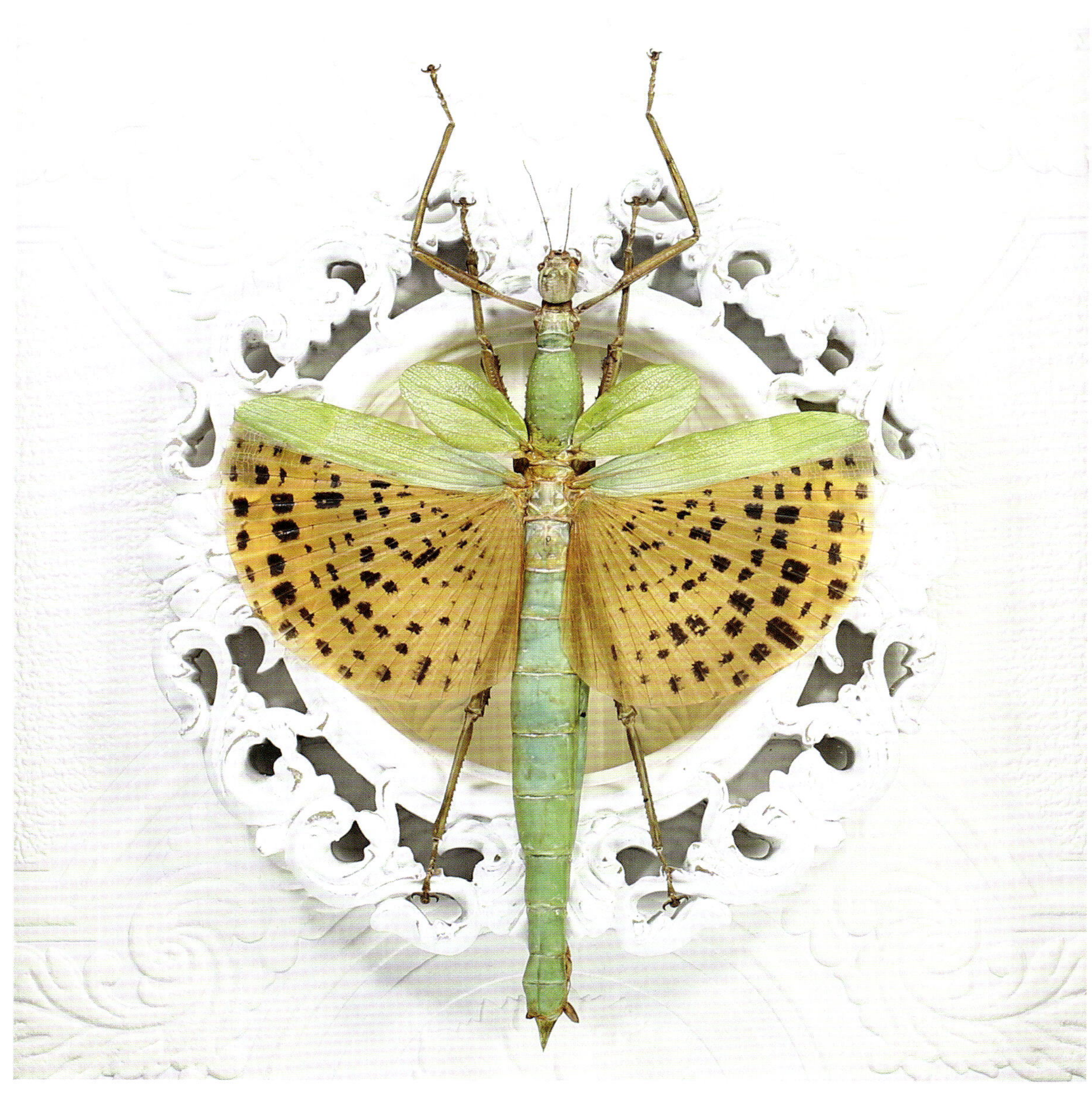

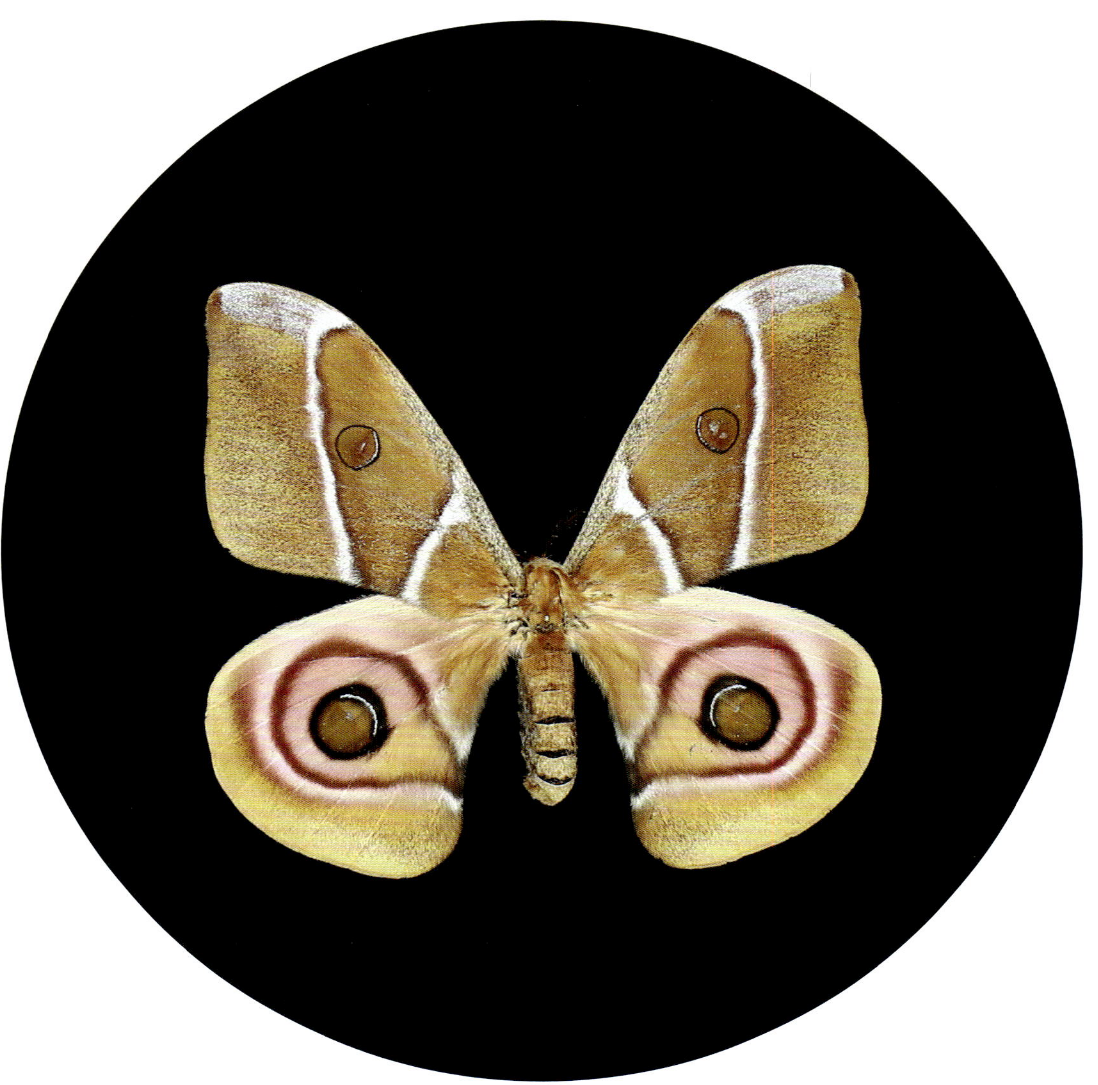

SURAKA SILKMOTH

Scientific Name
Antherina suraka

Geographic Range
Madagascar

In Madagascar, these distinctive moths are bred commercially, primarily for the purpose of producing silk. Remarkably, the fuzzy scales that cover their bodies can absorb up to 85 percent of the ultrasound emitted by echolocating bats. This adaptation reduces the moths' acoustic visibility, allowing them to evade detection by nocturnal predators that rely on sound to locate prey in the dark.

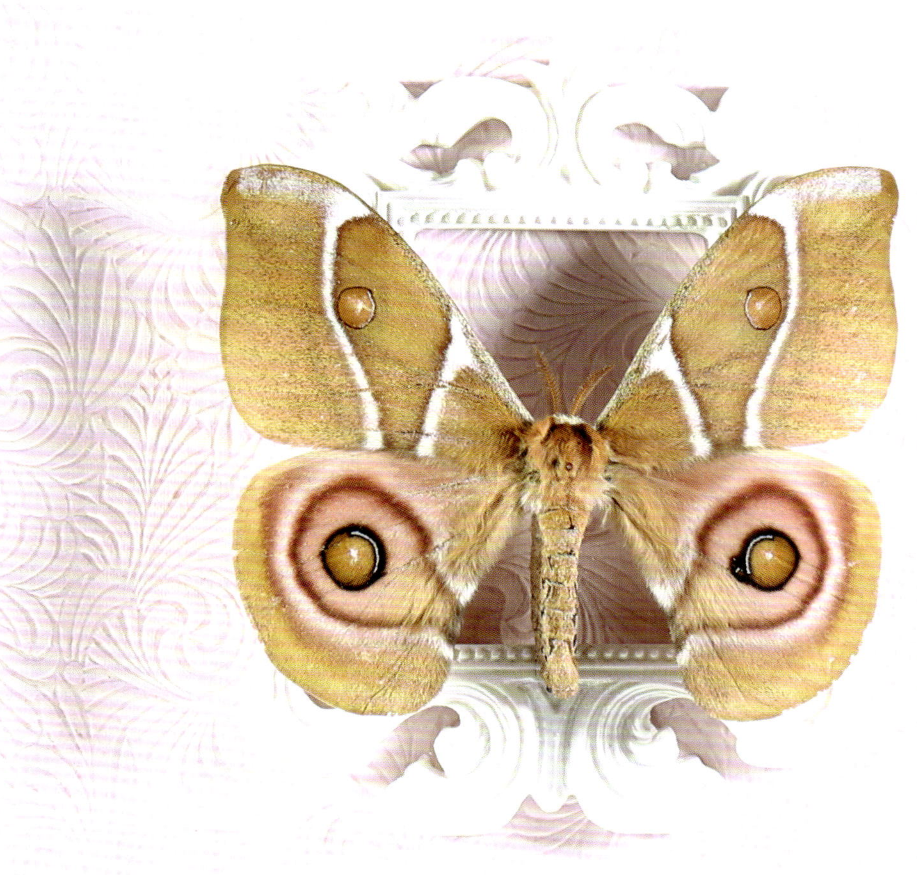

MONARCH

SCIENTIFIC NAME
Danaus plexippus

GEOGRAPHIC RANGE
North America, Central
America, northwestern
South America, Oceania,
and Europe

The well-known monarch butterfly migration from different parts of North America to central Mexico is an amazing multigenerational journey in which individual monarchs complete only a portion of the entire migration route, relying on following generations to continue the trek. These incredible creatures can fly 2,500 miles (4,023 kilometers) or even more, despite the fact that most live for just a few weeks. They rely heavily on their finely tuned ability to navigate using the sun and the earth's magnetic field; their specialized antennae help them determine the time of day.

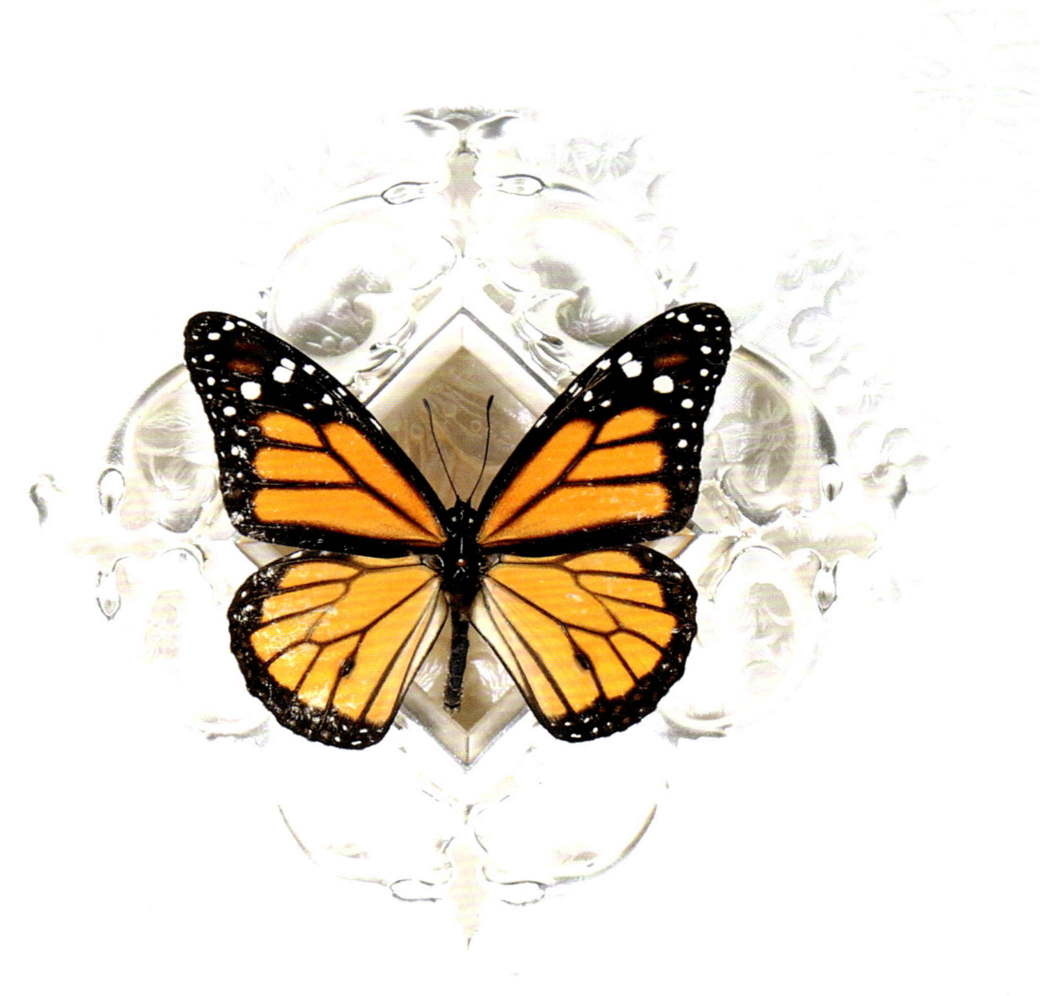

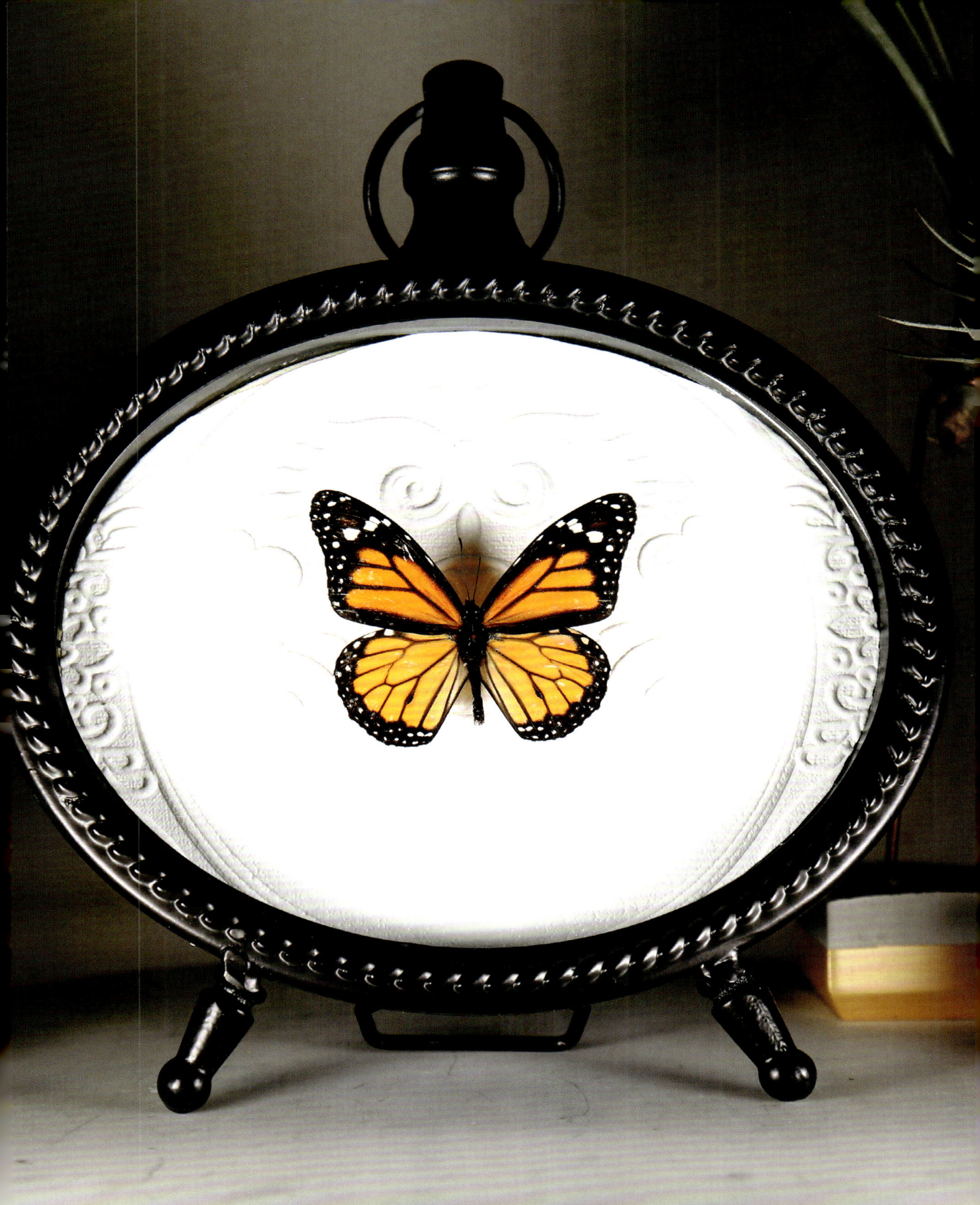

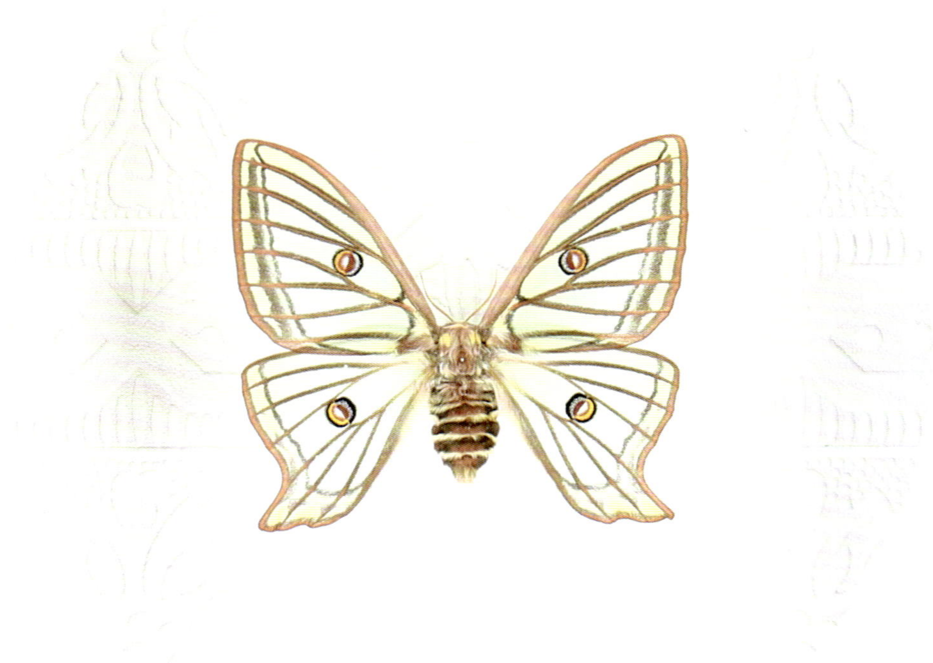

SPANISH MOON MOTH

SCIENTIFIC NAME
Graellsia isabellae

GEOGRAPHIC RANGE
Spain, France, and
Switzerland

The moon moth inhabits the rugged and often challenging highlands of the Pyrenees, as well as other similar mountainous regions where the climate is cold and harsh. Considered a relic species that dates back to the Ice Age or even earlier, this moth has persisted in its unique habitat due to thousands of years of environmental consistency, becoming a staple of the region's natural history.

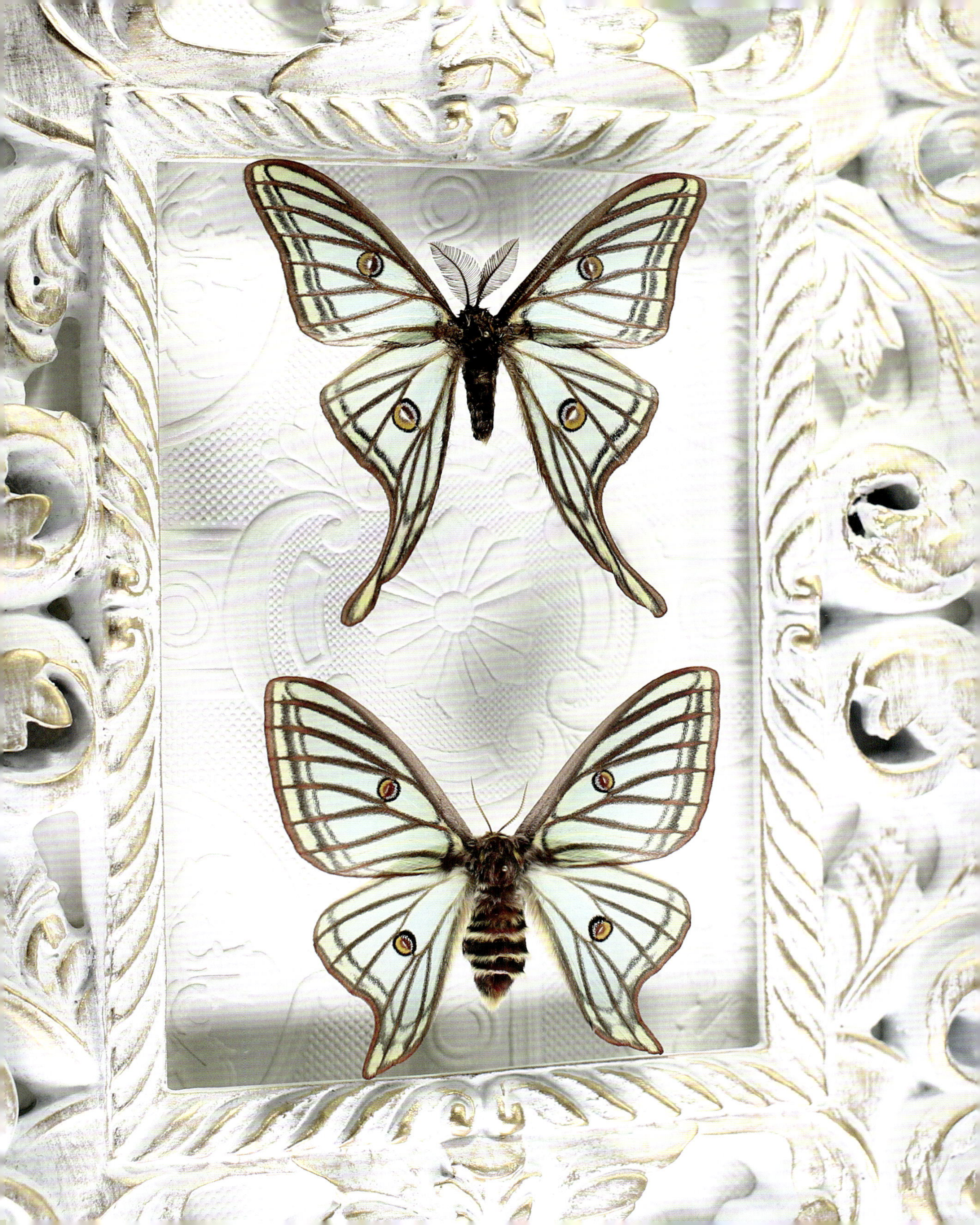

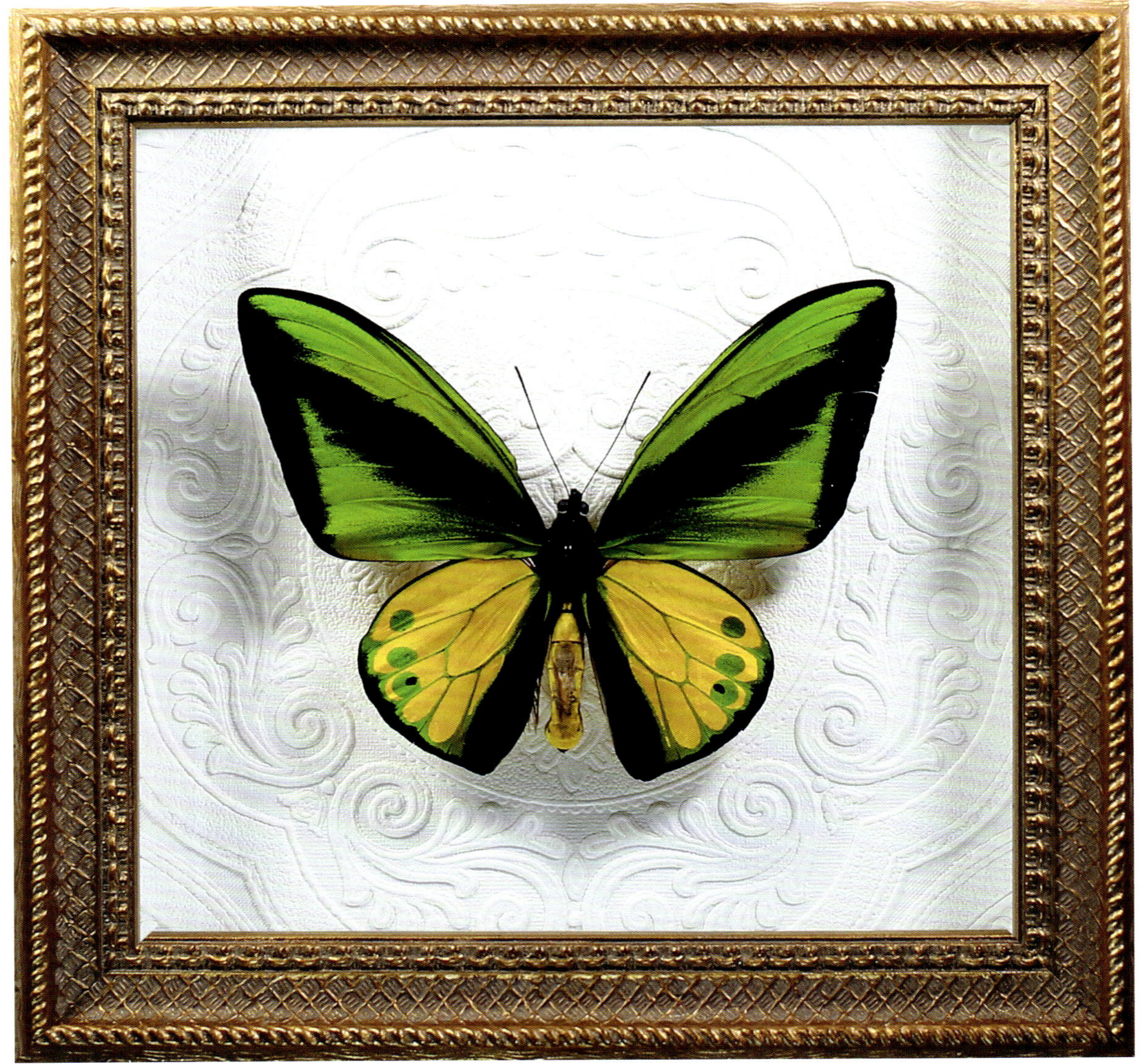

GOLIATH BIRDWING

Scientific Name
Ornithoptera goliath

Geographic Range
Moluccas and New Guinea

With a wingspan of up to 11 inches (28 centimeters), the goliath is the second-largest butterfly in the world, following the Queen Alexandra's birdwing. Its namesake is the biblical giant Goliath, who was defeated by David in an epic battle.

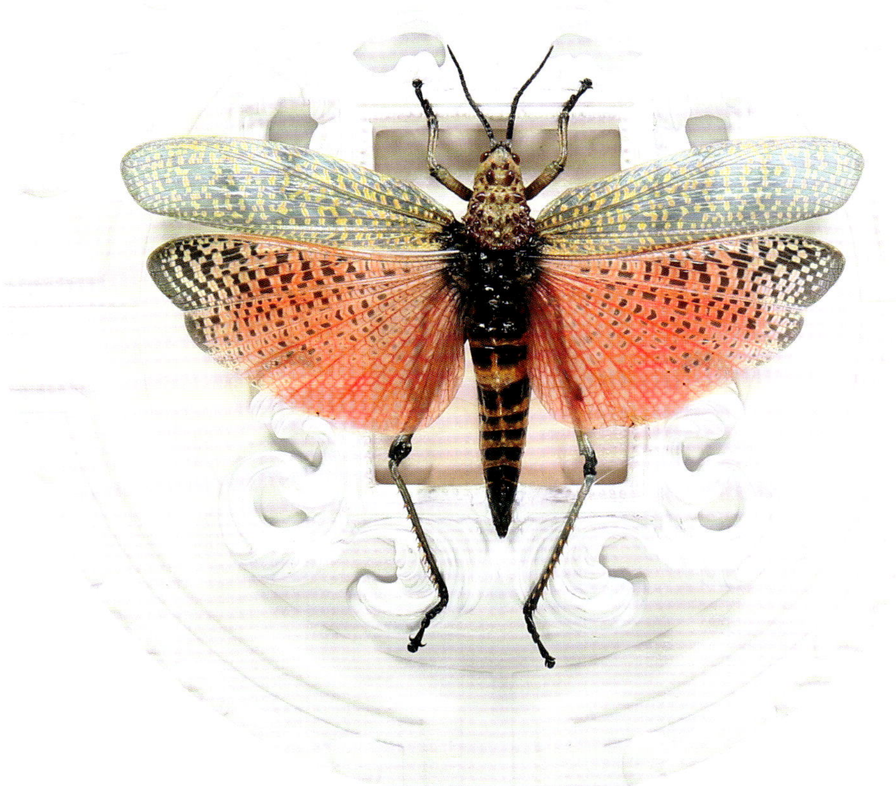

RAINBOW MILKWEED LOCUST

Scientific Name
Phymateus saxosus

Geographic Range
Madagascar

When threatened, these locusts release a highly poisonous froth from specialized openings called spiracles, located beneath their wings. The locust's toxicity is derived from the milkweed plant that it consumes.

ATLAS MOTH

SCIENTIFIC NAME
Attacus atlas

GEOGRAPHIC RANGE
Asia

Atlas moths boast impressive wingspans that can reach almost 1 foot (30.5 centimeters) in length. To deter predators, these creatures drop to the ground and wriggle their large, showy wings in a manner that closely imitates the appearance of a slithering snake.

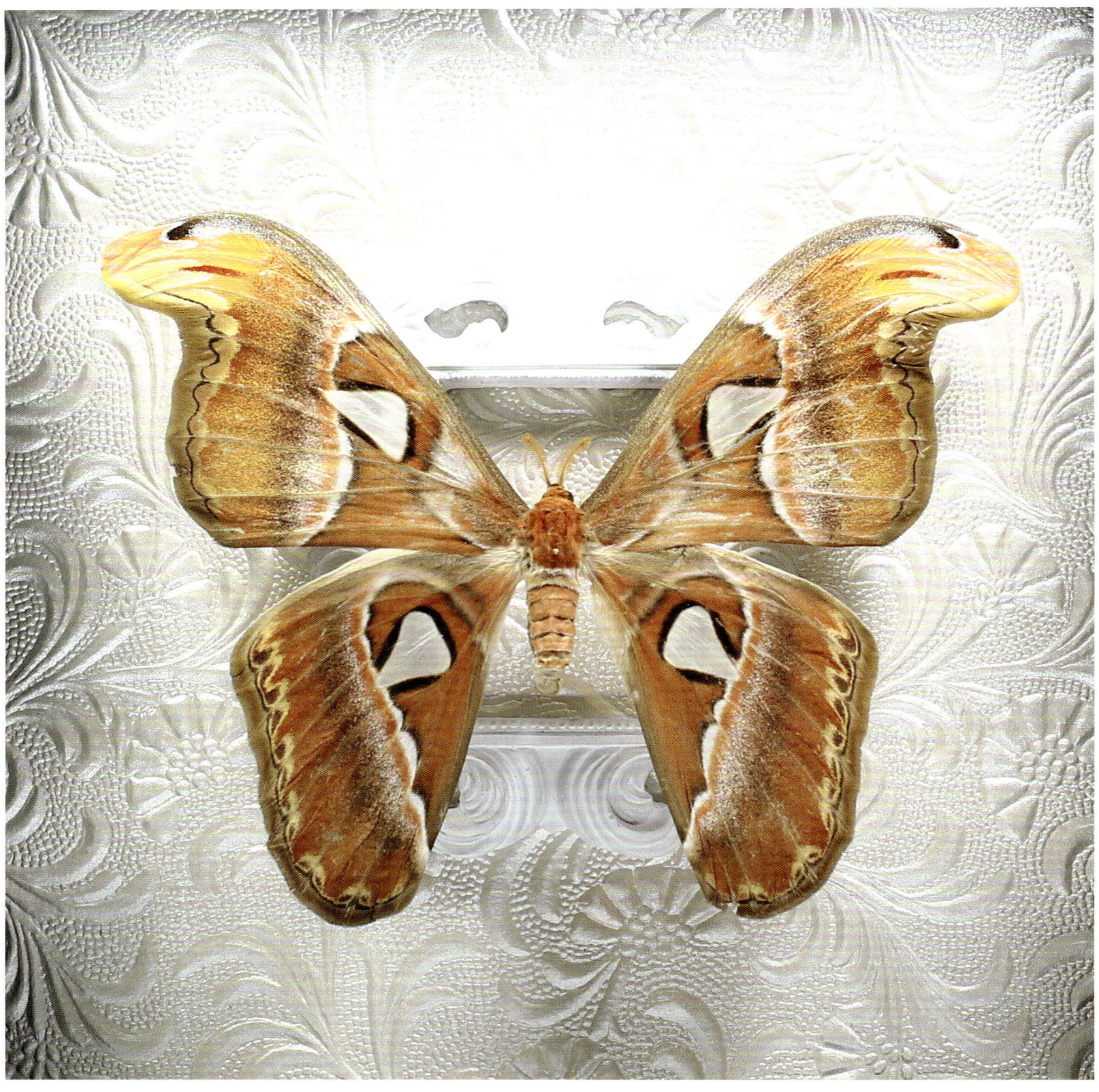

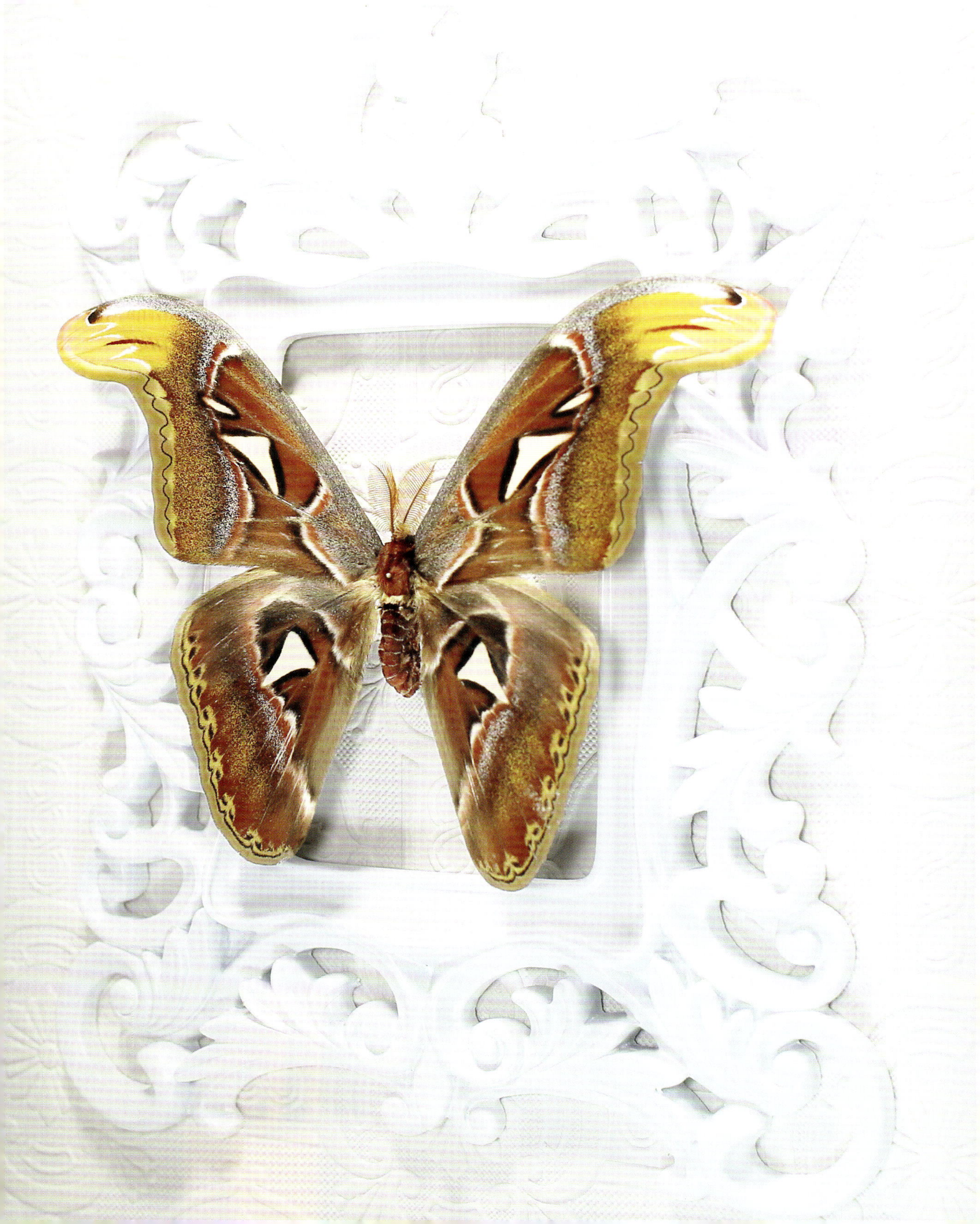

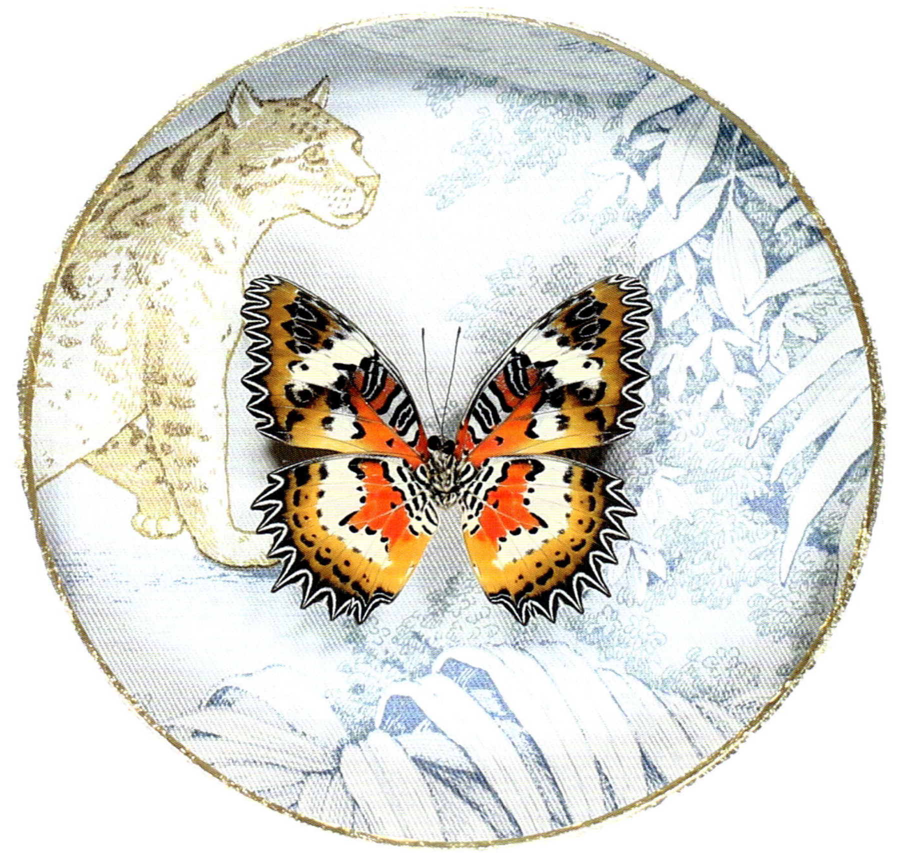

RED LACEWING

SCIENTIFIC NAME
Cethosia biblis

GEOGRAPHIC RANGE
Asia

In males, the intense red of the dorsal side of this species' wings serves as a clear warning to potential predators. Lacewing caterpillars eat poisonous passion vines, making them unpalatable and providing an added layer of protection in their natural environment.

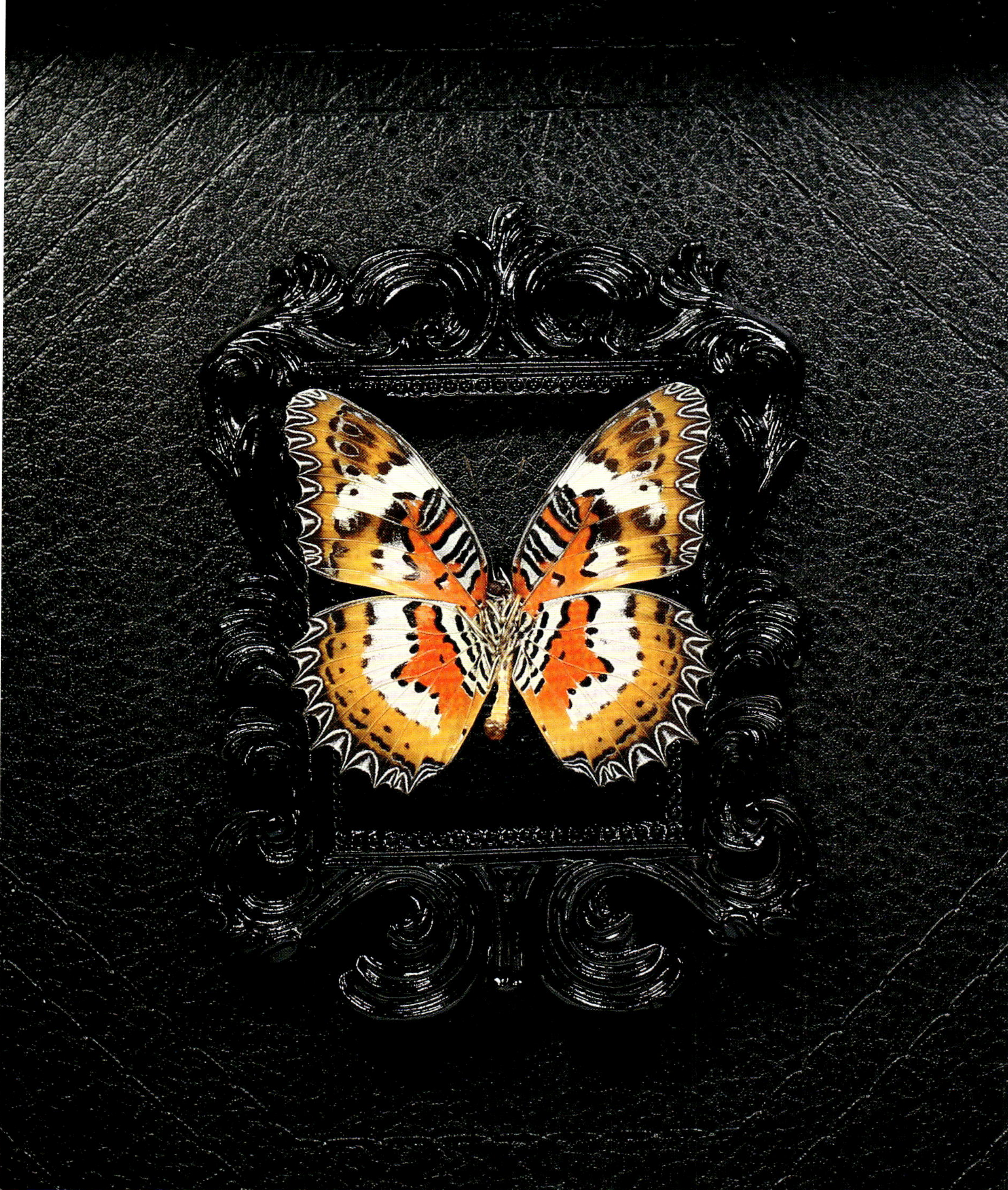

OWL MOTH

Scientific Name
Brahmaea wallichii

Geographic Range
India, Nepal, Bhutan,
Myanmar, China,
Taiwan, and Japan

Owl moths are primarily active during the night; during the day, they often rest with outspread wings on tree trunks or on the ground below. Their wings are intricately patterned with distinct eyespots that resemble the face of an owl. When disturbed or approached by potential threats, these moths do not simply fly away; they react by fiercely shaking their bodies as a warning.

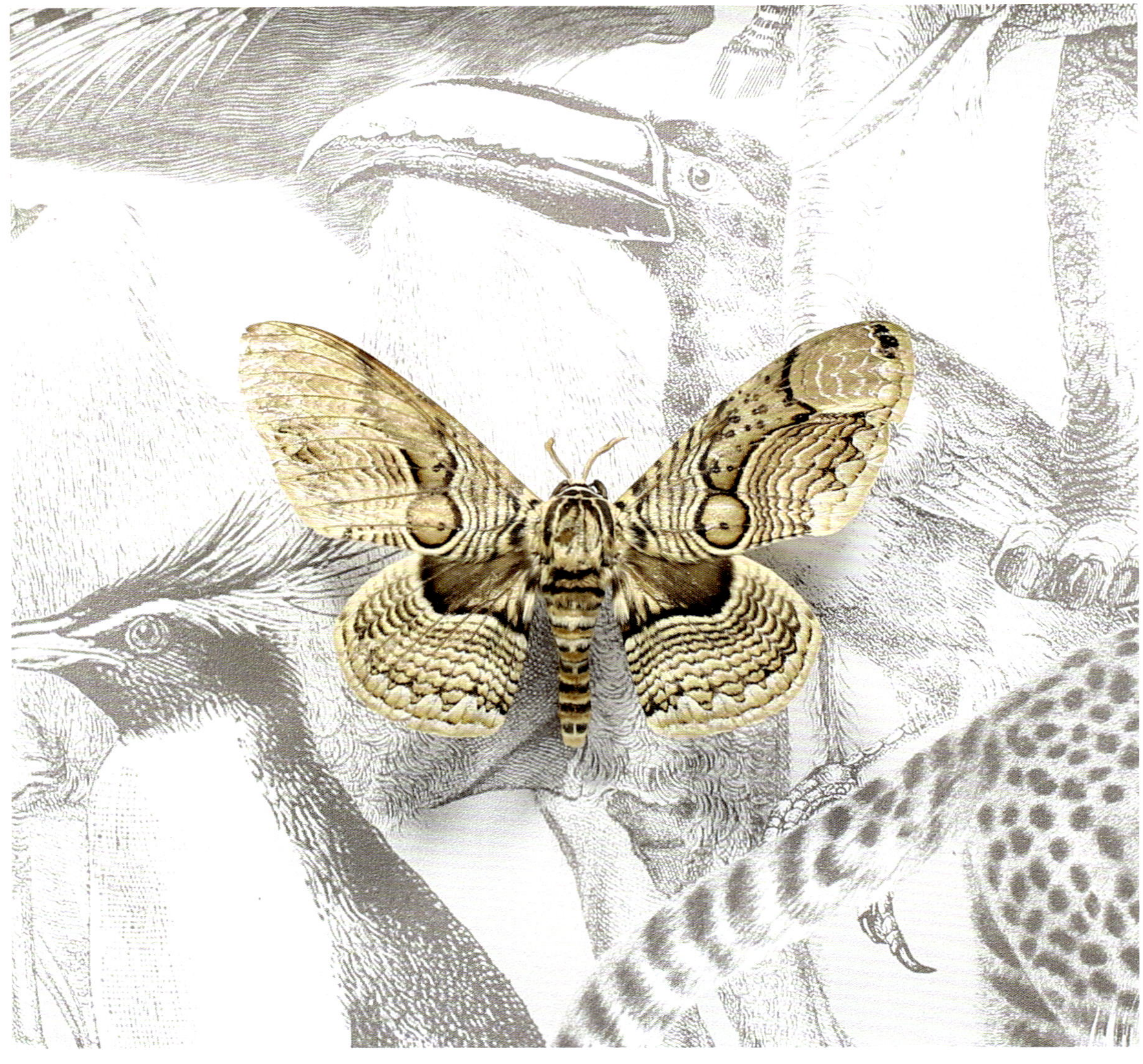

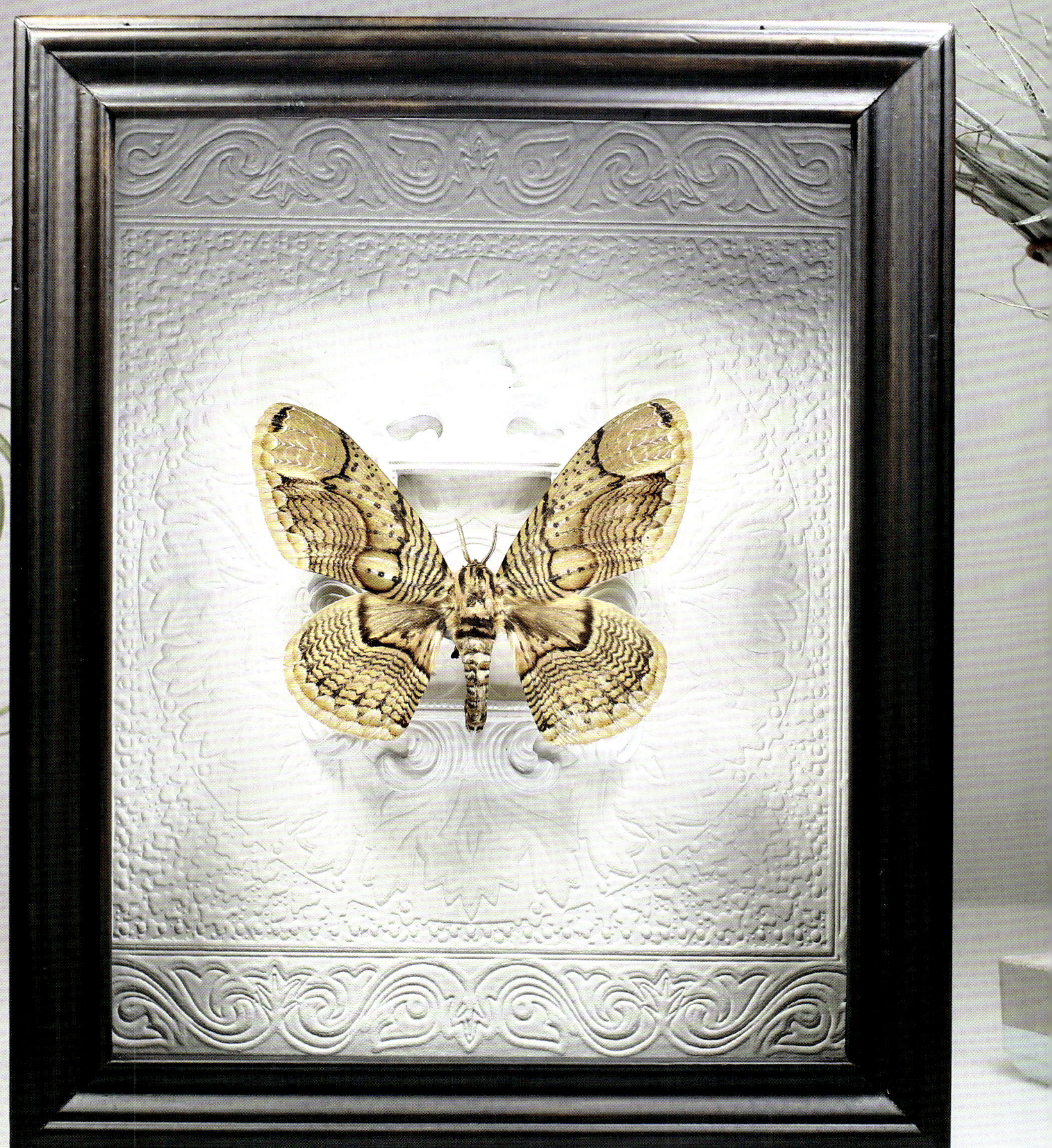

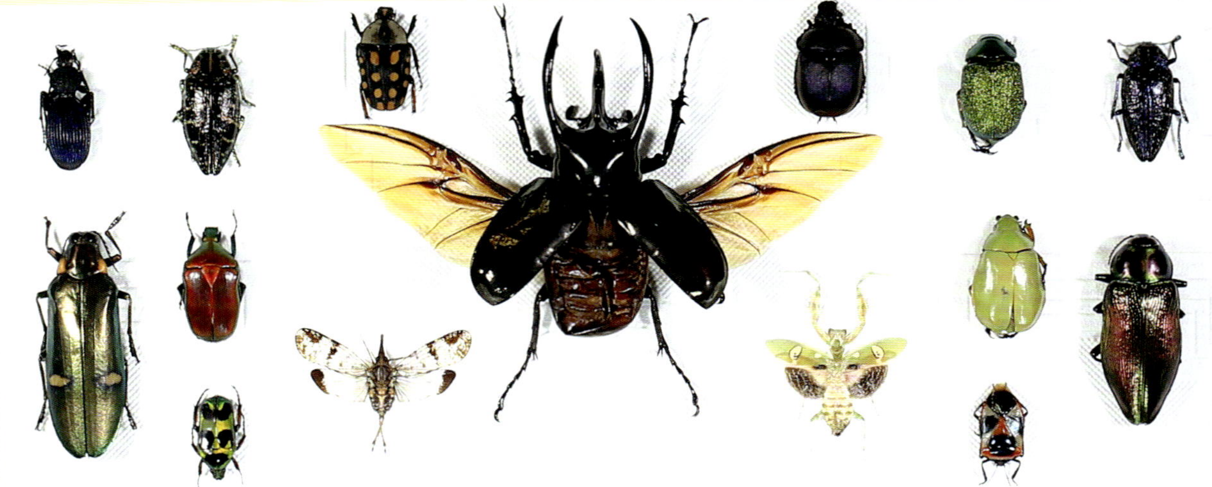

THE GEOMETRY of SURVIVAL

Insect Shape

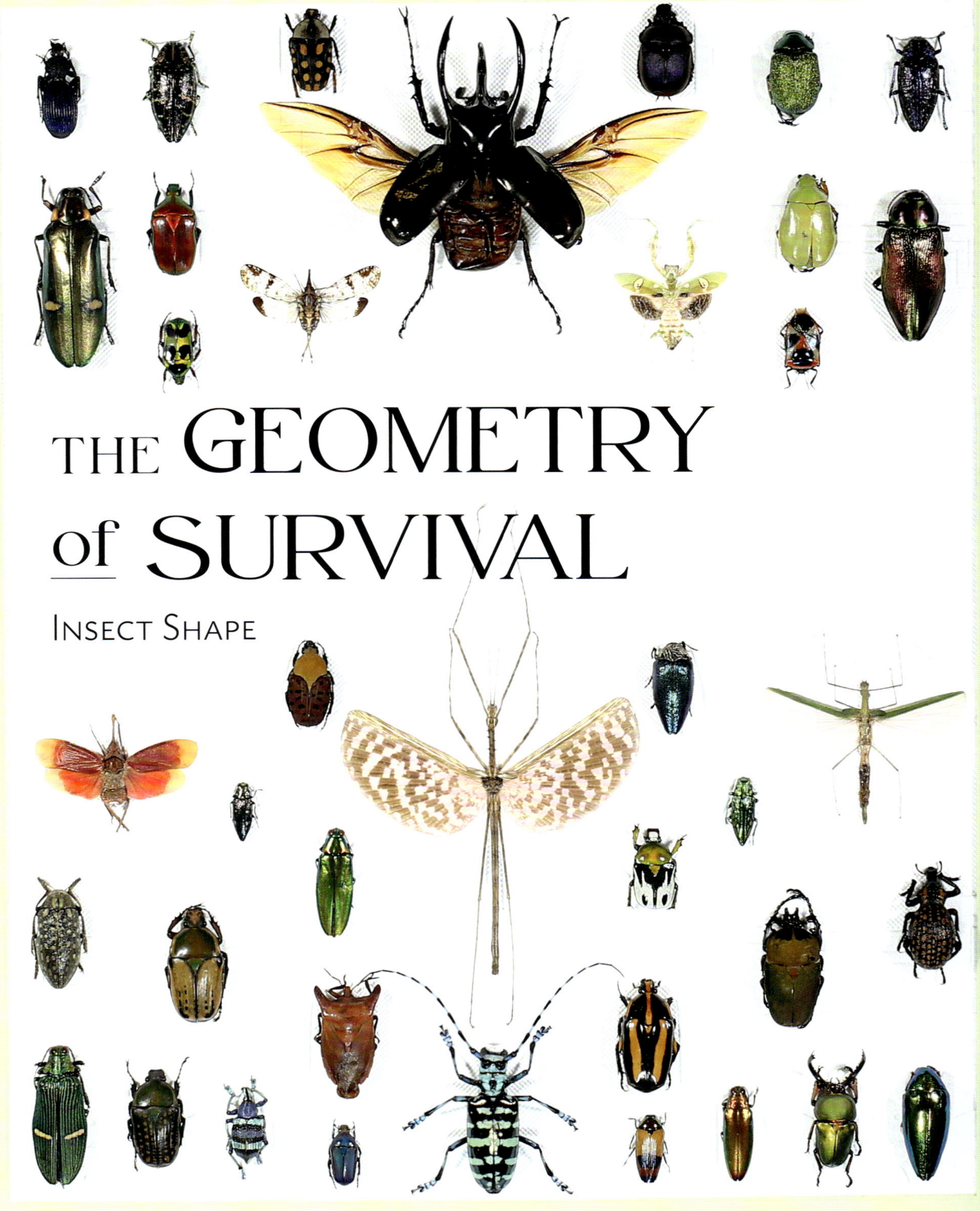

AN INSECT'S SHAPE ACTS AS AN ARCHITECTURAL BLUEPRINT THAT DEFINES its every movement, every function, and every interaction. The extraordinary diversity observed in insect shapes is a direct reflection of the countless pressures they have faced over millennia.

Some insects are built for speed. The long, slender bodies and powerful legs of grasshoppers and ants enable them to zip across the landscape in search of food or suitable shelter. Other insects are designed for stealth. Flat, disk-shaped beetles and leaf-mimicking bugs blend expertly into their surroundings, using their unique shapes to hide in plain sight from potential predators. Meanwhile, the forms of vital pollinators like bees and butterflies are optimized for hovering, feeding, and carrying pollen from one flower to another. The elongated bodies of wasps and the agile figures of praying mantises are sculpted for precision and predation, allowing them to become proficient hunters in their respective habitats. Some insects have developed sharp spikes, hardened shells, or robust armored exoskeletons that make them nearly invulnerable to a wide array of hungry predators.

The insect world reveals a breathtaking array of forms—from the sleek to the spiny, from the compact to the elongated—each one a masterstroke in the complex geometry of survival.

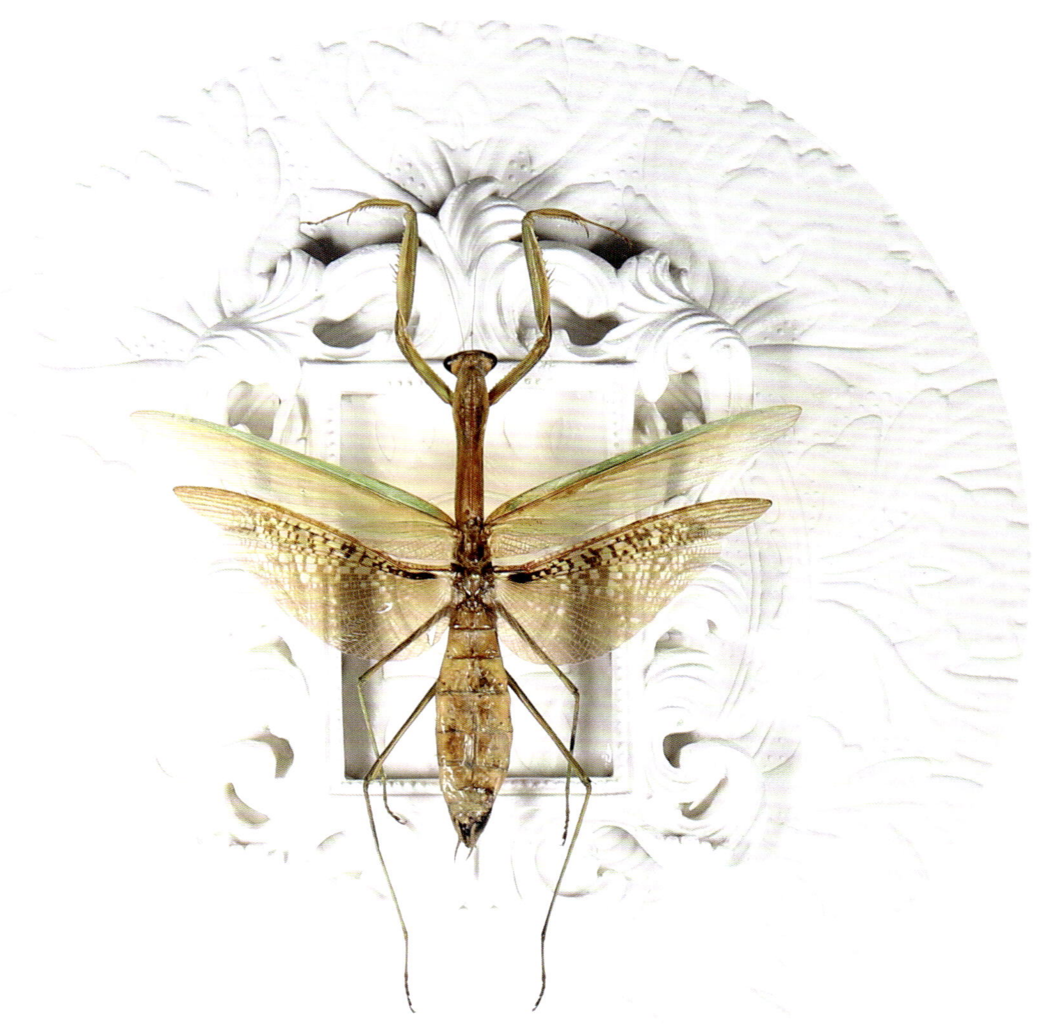

CHINESE MANTIS

SCIENTIFIC NAME
Tenodera sinensis

GEOGRAPHIC RANGE
South Asia, East Asia,
Southeast Asia, Micronesia,
Hawaii, and North America

Two distinct martial arts styles have been specifically developed to mimic the fluid and agile movements of the Chinese mantis: praying mantis kung fu and southern praying mantis kung fu. These styles include fast, precise hand techniques inspired by the movement of mantis claws.

PEANUT-HEADED LANTERNFLY

This insect possesses a unique structure that grows from its head. When viewed from directly above, this unusual feature resembles a peanut. However, when seen from the side, it resembles the silhouette of a lizard. This interesting appendage helps the peanut-headed bug camouflage.

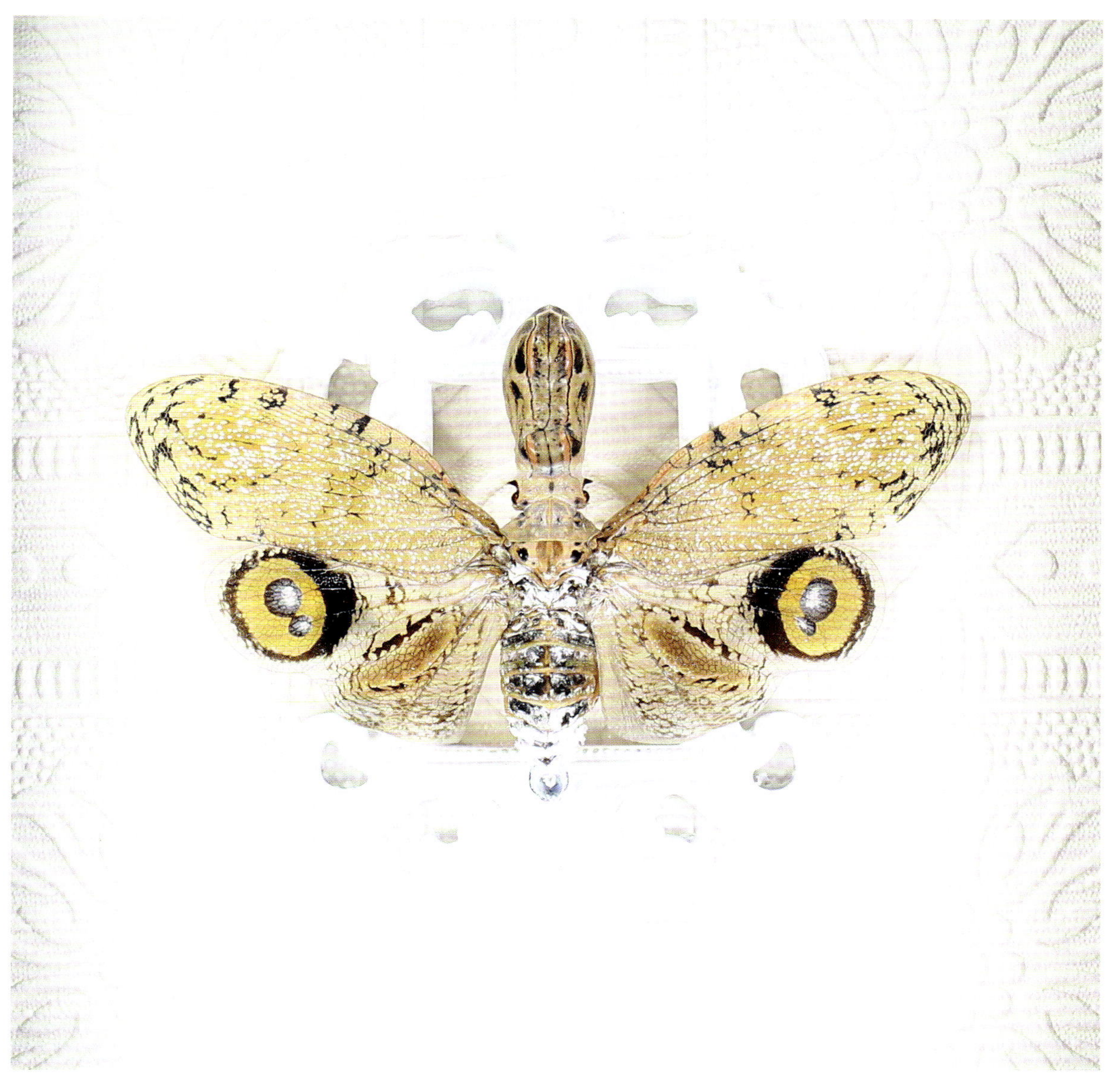

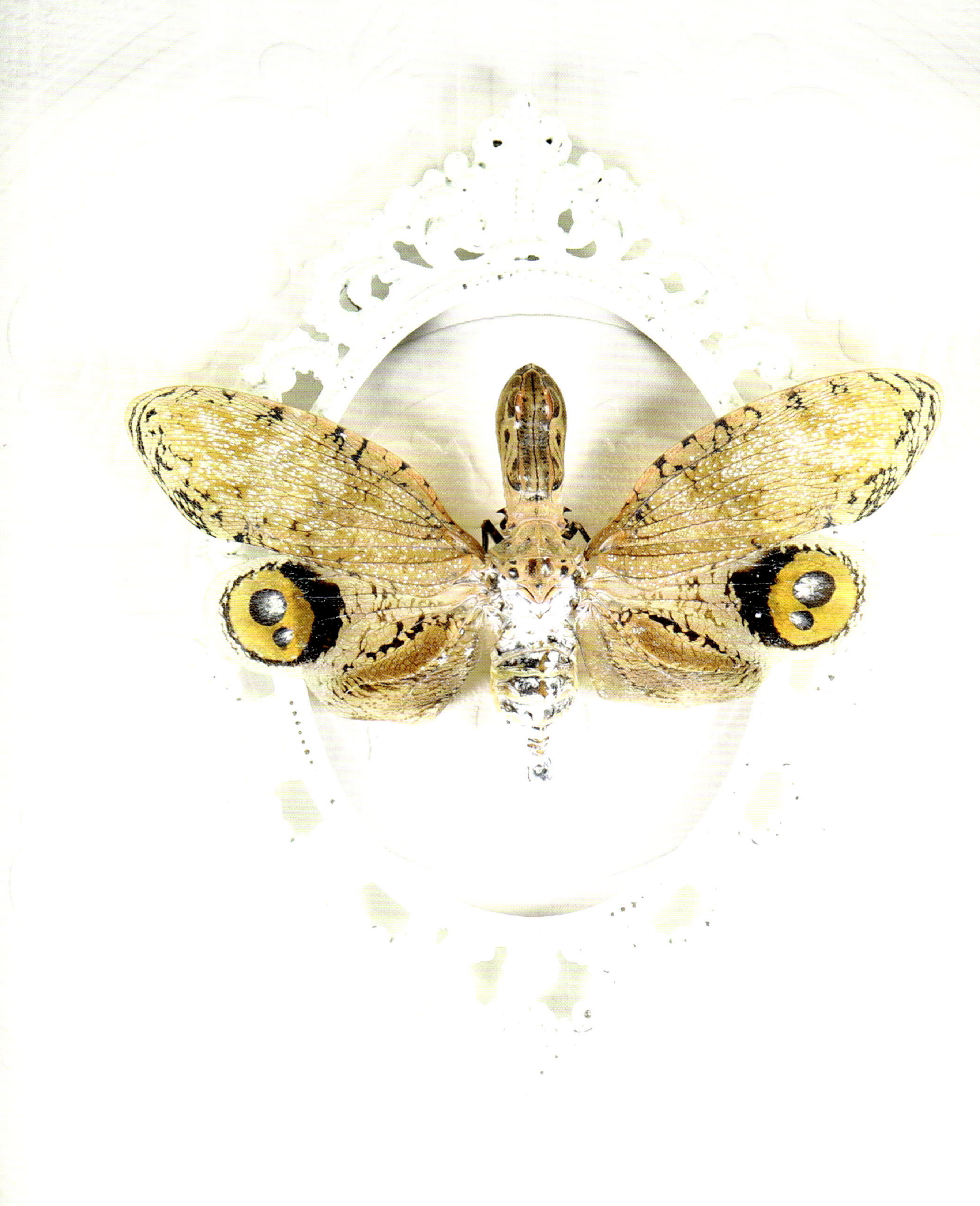

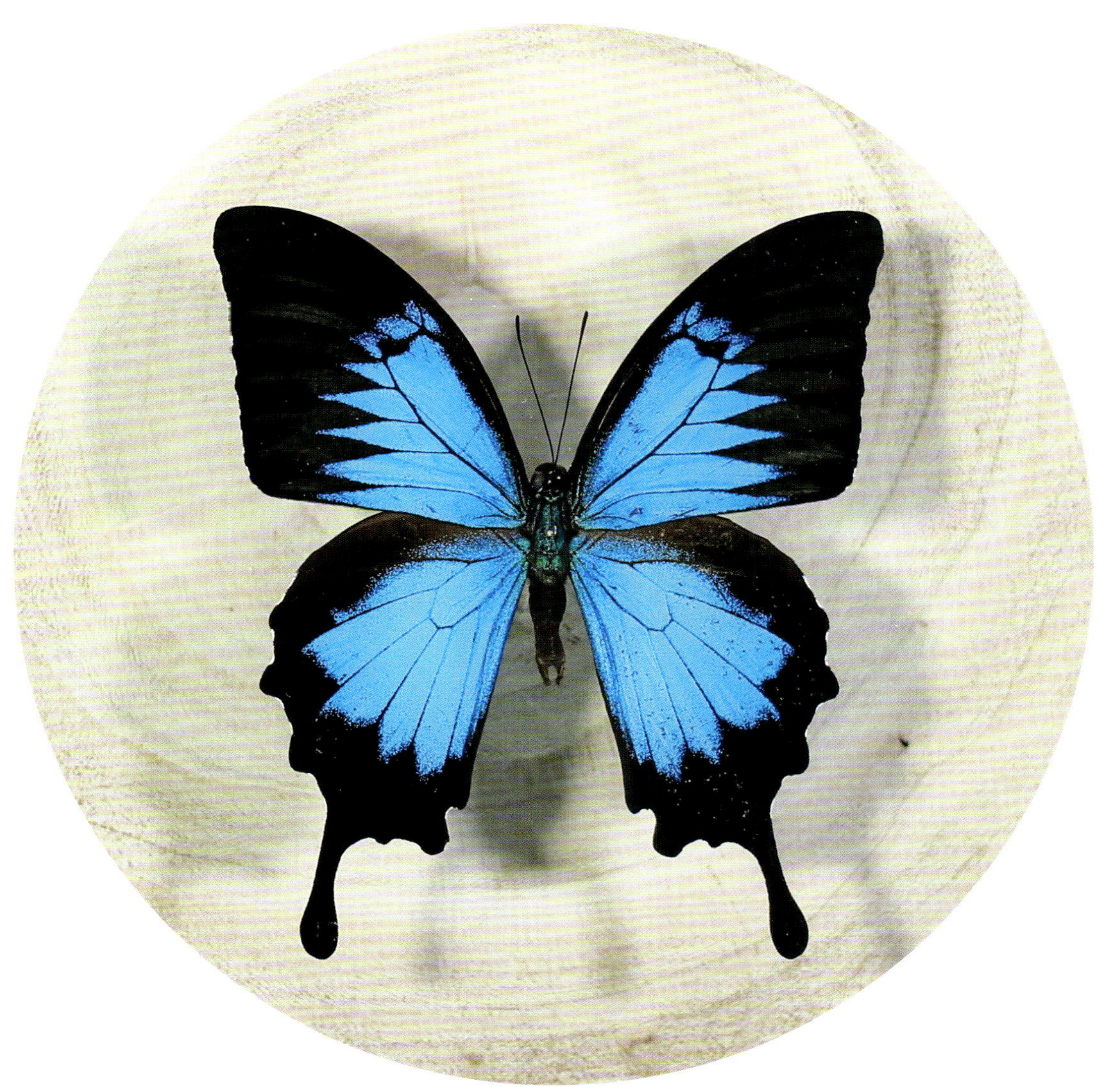

ULYSSES BUTTERFLY

SCIENTIFIC NAME
Papilio ulysses

GEOGRAPHIC RANGE
Indonesia, Papua New
Guinea, Solomon Islands,
and Australia

Ulysses butterflies have rapid, erratic flight patterns, making them difficult to observe in the wild. The males of this species are attracted to the color blue and often mistake blue objects for potential mates.

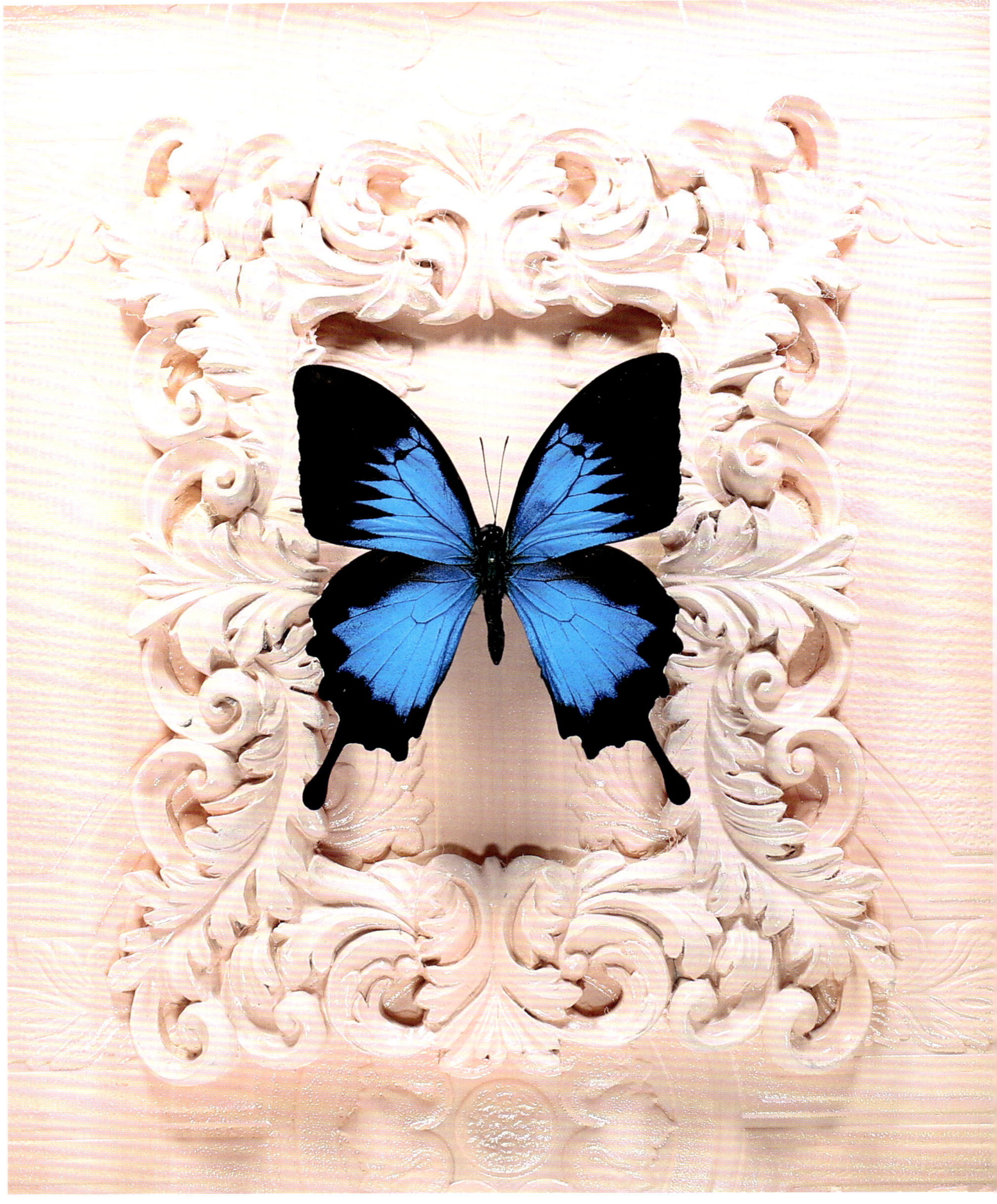

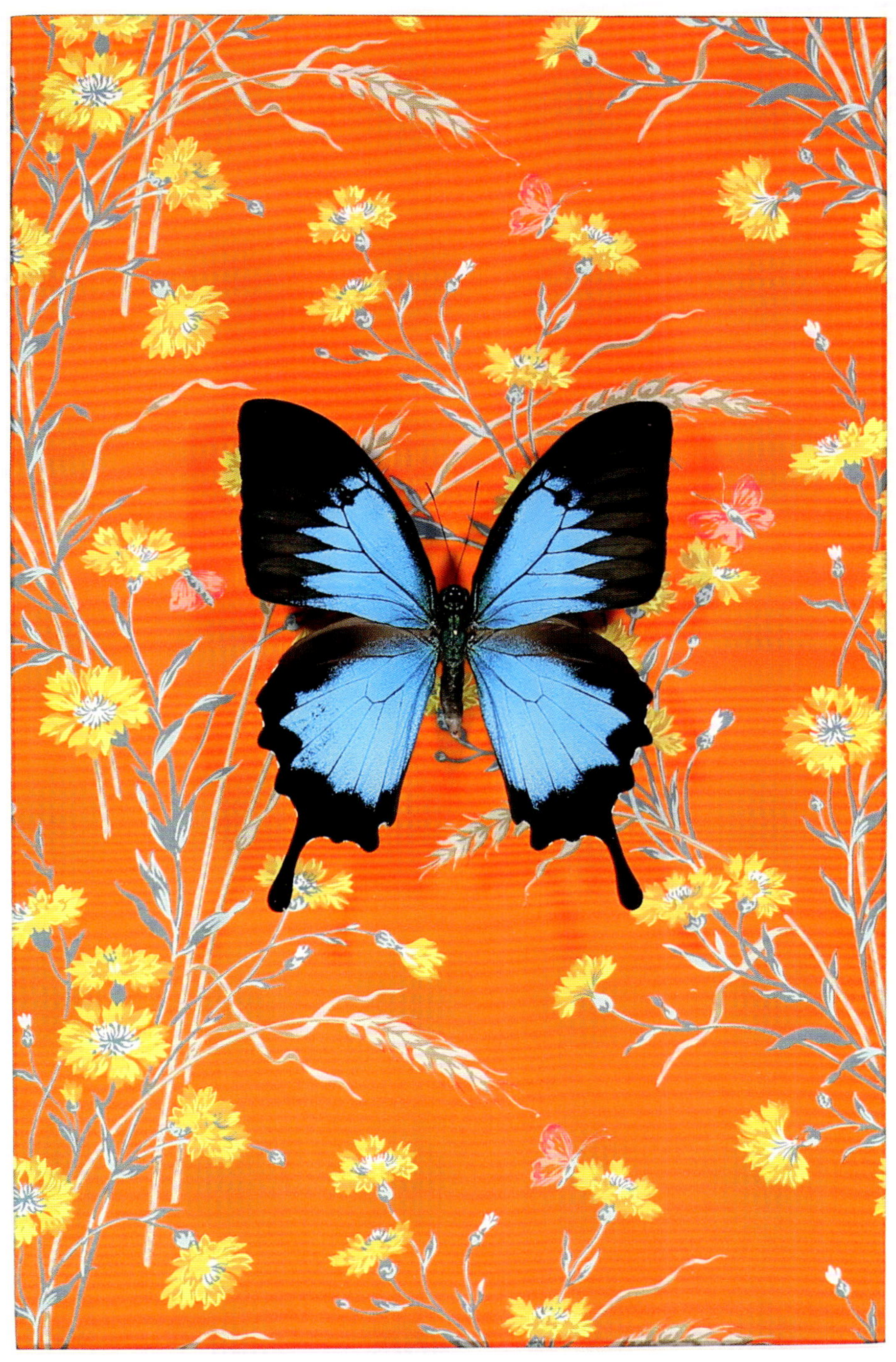

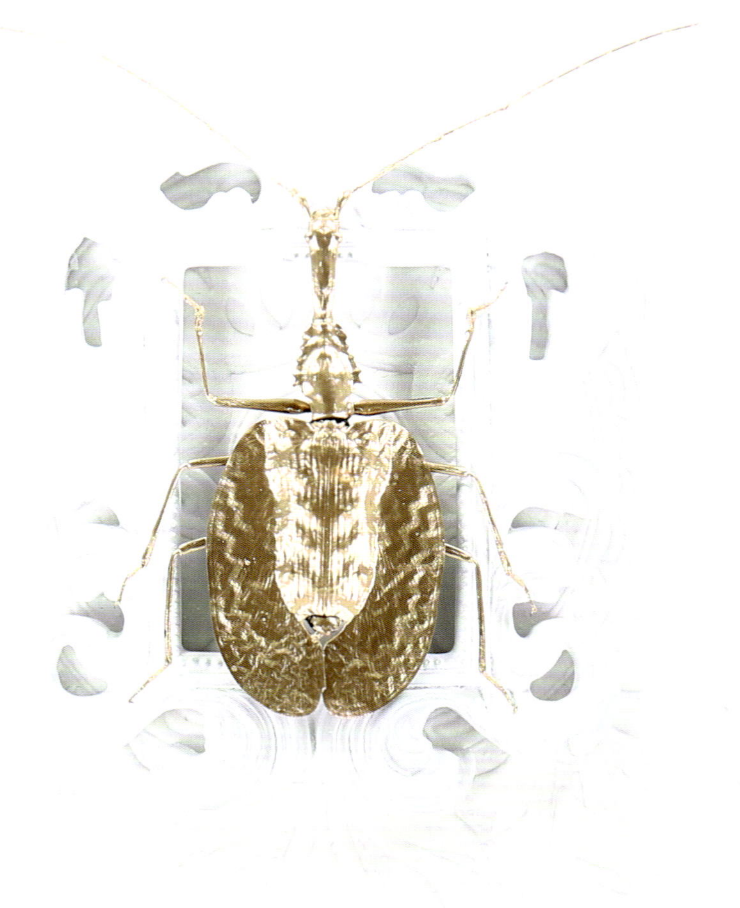

VIOLIN BEETLE

SCIENTIFIC NAME
Mormolyce phyllodes

GEOGRAPHIC RANGE
Thailand, Malaysia,
Brunei, and Indonesia

The hardened forewings of the violin beetle are extremely flat and expand laterally to allow it to maneuver between tight layers of thick bark and fungus.

FIGHTING GIANT STAG BEETLE

SCIENTIFIC NAME
Hexarthrius parryi

GEOGRAPHIC RANGE
Southern China and
Southeast Asia

These beetles use their menacing jaws to engage in intense wrestling matches with other males, usually to determine who gets to mate with a female.

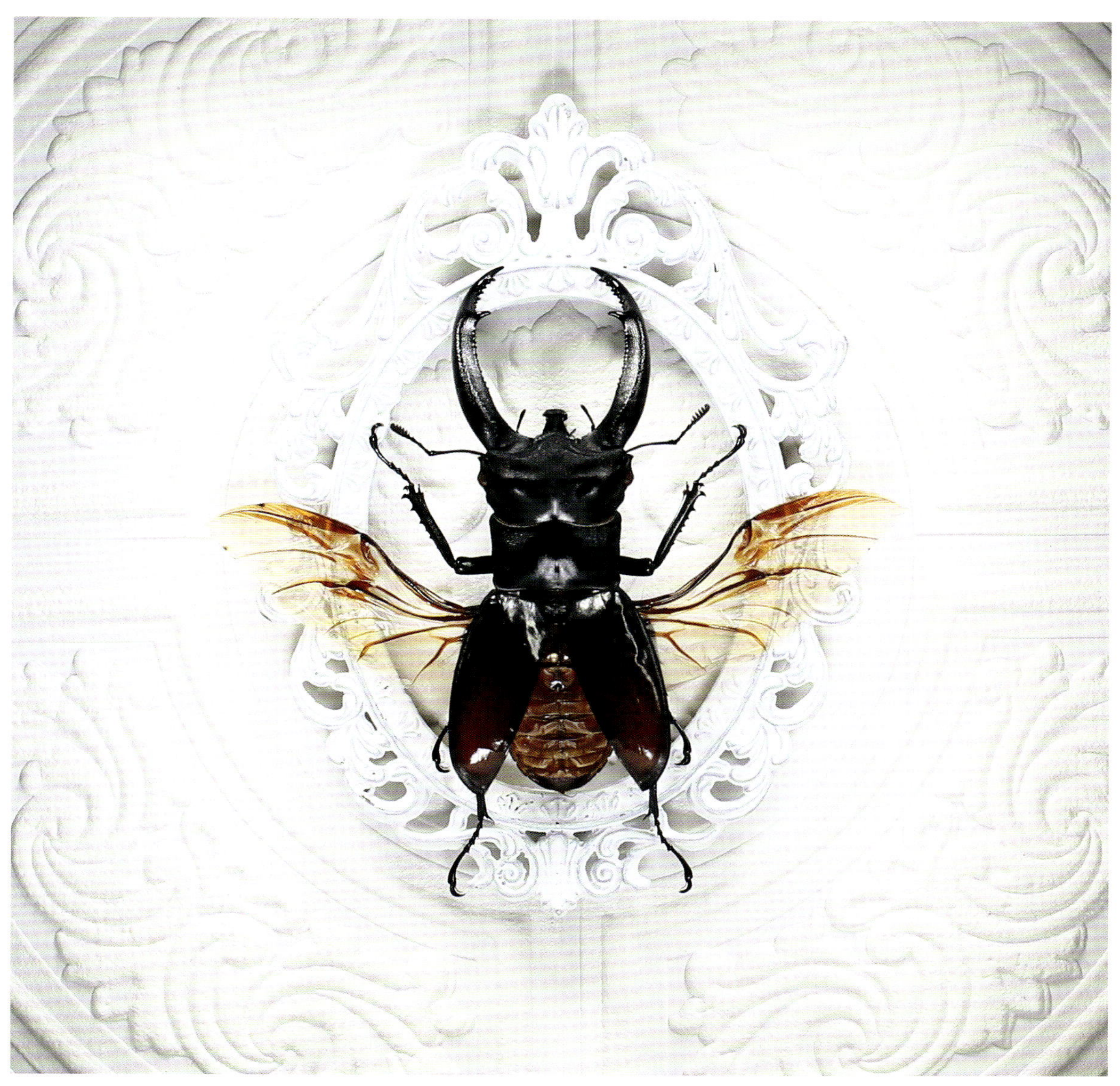

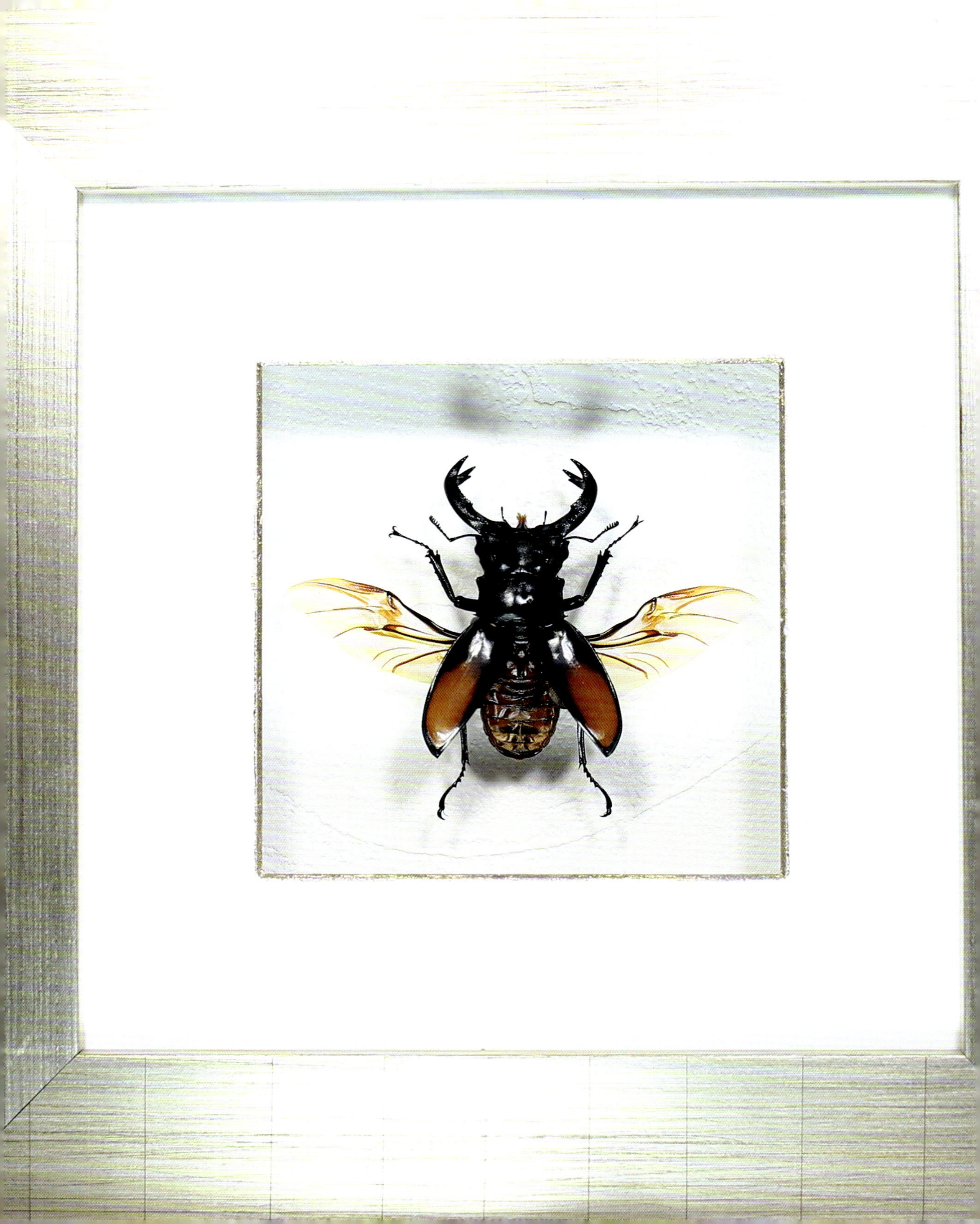

GIANT KATYDID

SCIENTIFIC NAME
Pseudophyllus hercules

GEOGRAPHIC RANGE
Malaysia, Brunei, and
Indonesia

The genus name *Pseudophyllus* comes from the ancient Greek for "false" and "leaf," speaking to this creature's morphological ability to mimic leaves.

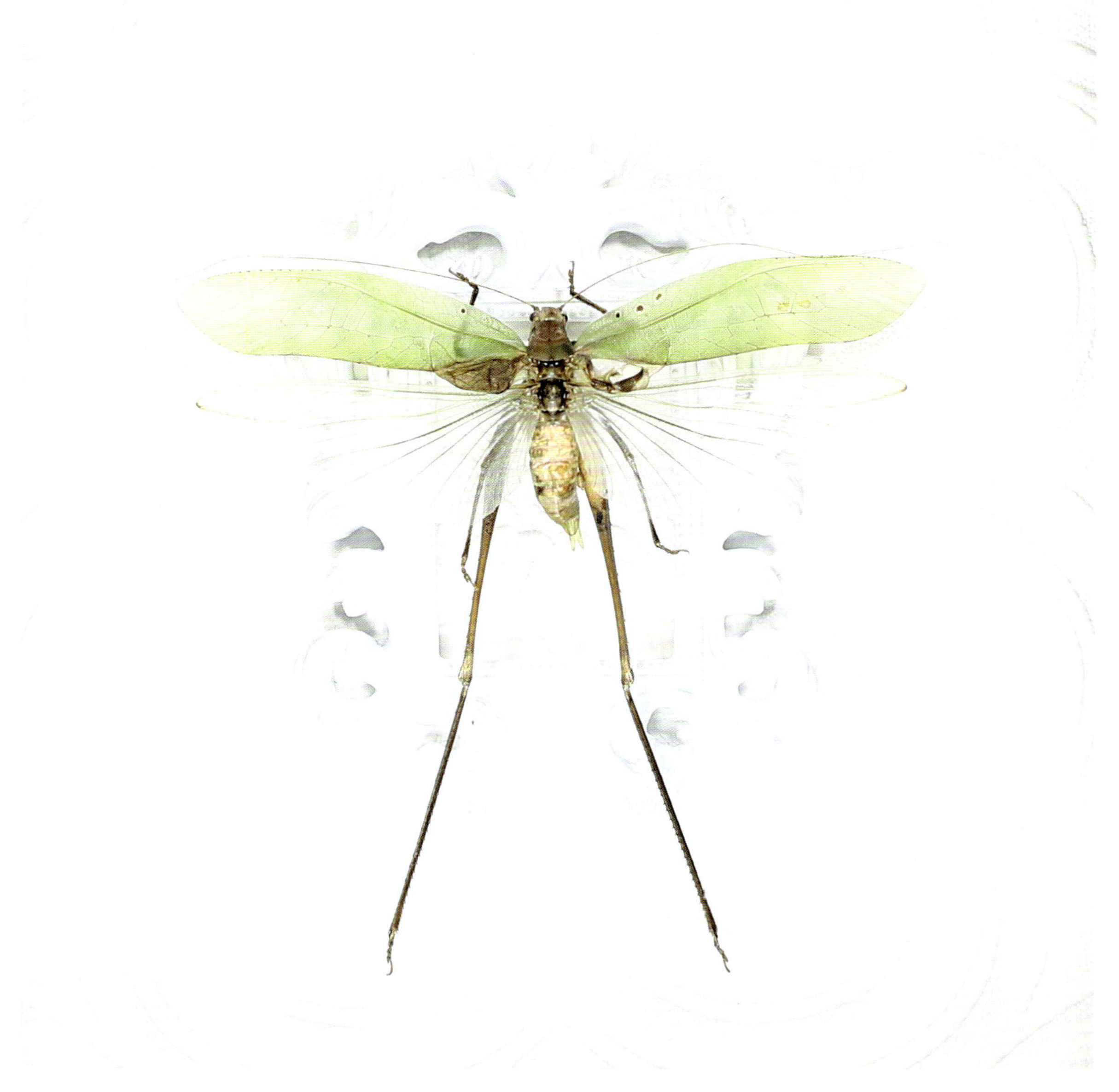

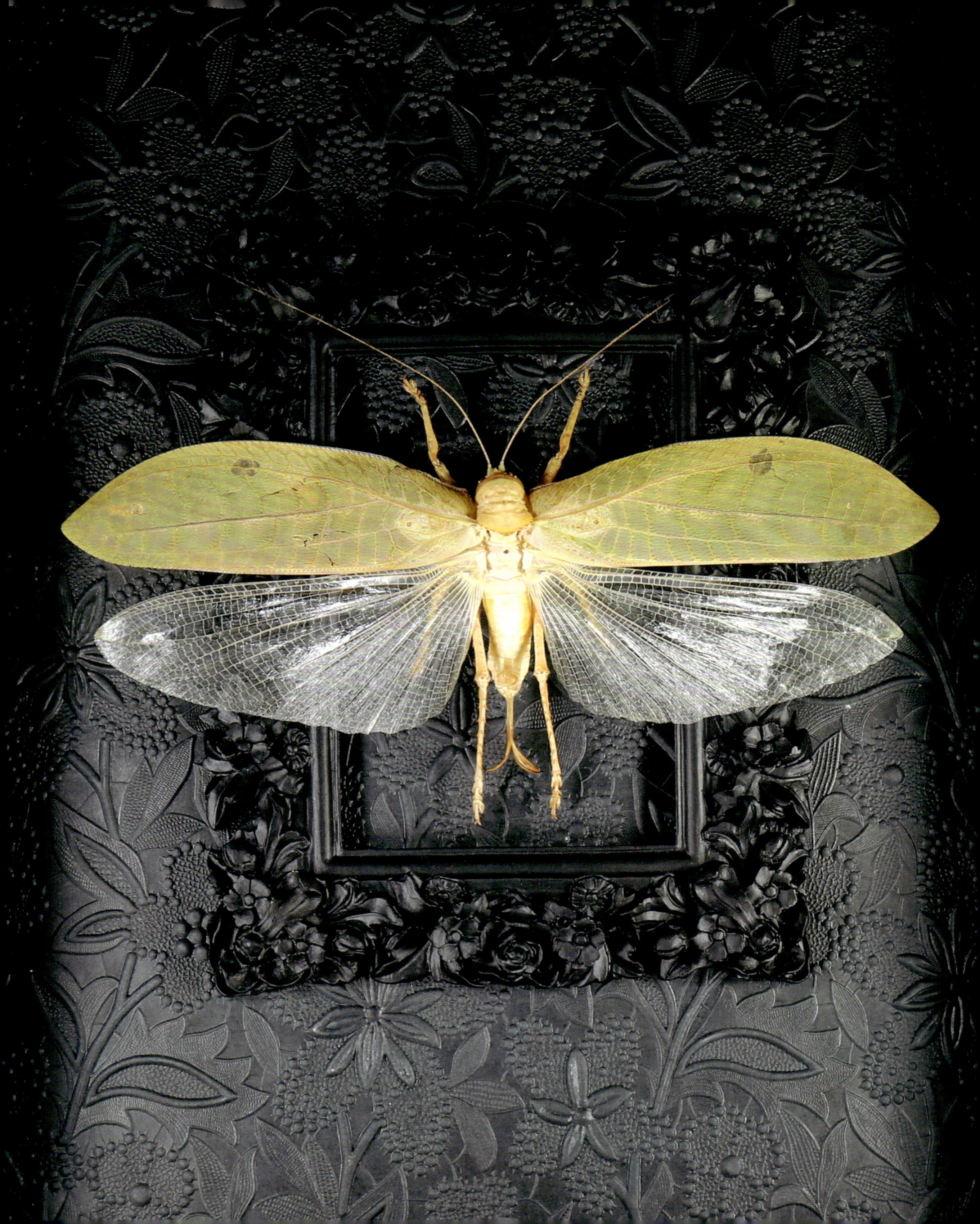

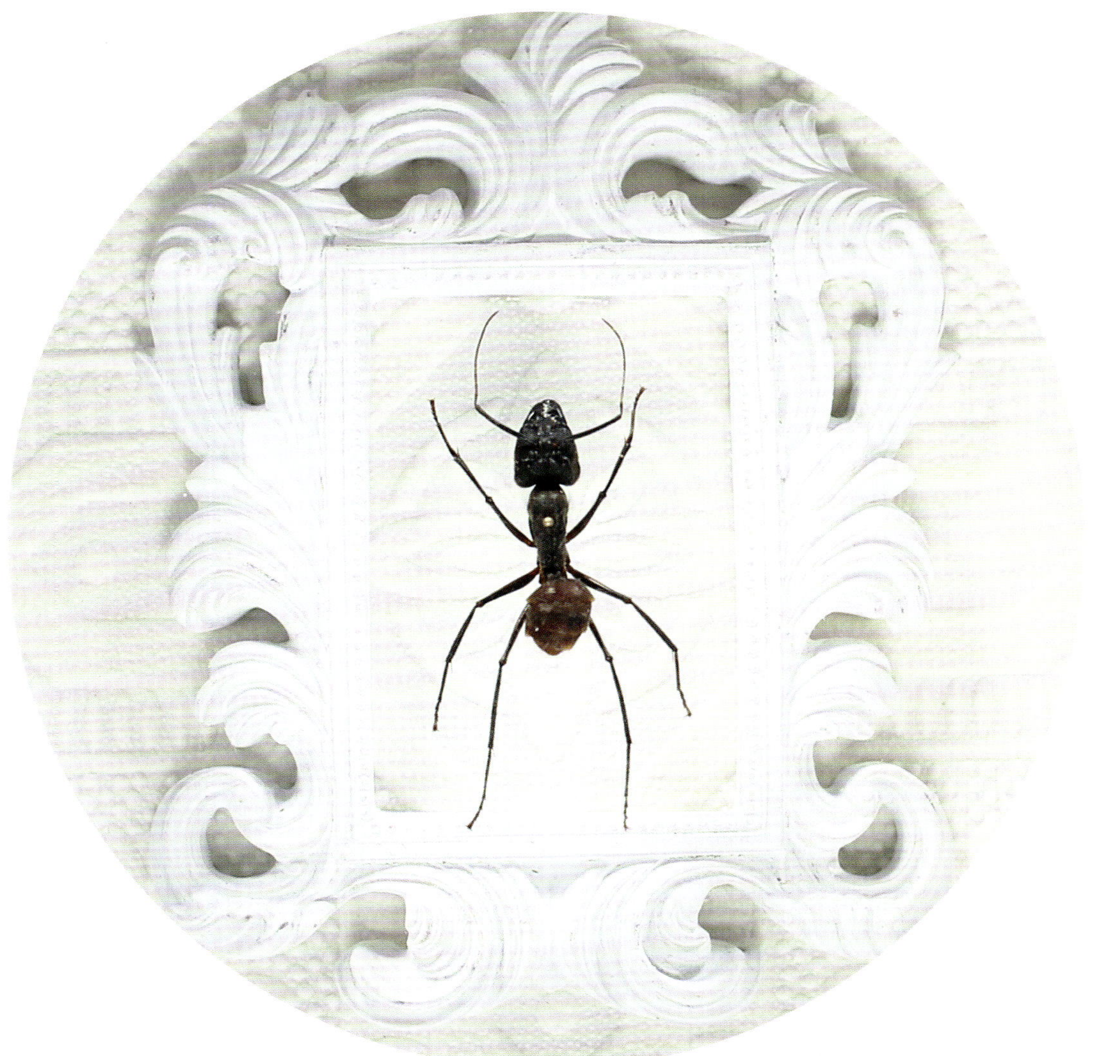

BULLET ANT

SCIENTIFIC NAME
Paraponera clavata

GEOGRAPHIC RANGE
Central America and
South America

The bullet ant is famous for its namesake sting, often described as one of the most painful in the insect world. The pain can last an entire day, earning the creature the nickname "24-hour ant." In one part of Brazil, the Sateré-Mawé people incorporate these ants into coming-of-age rituals in which enduring multiple stings is a test of strength and resilience.

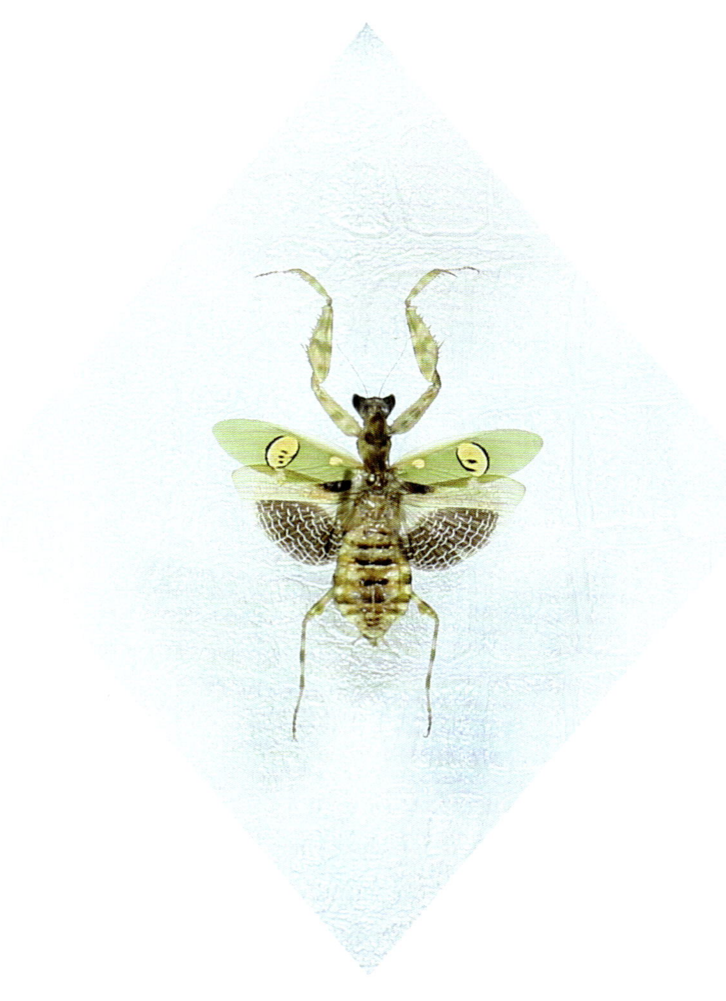

JEWELED FLOWER MANTIS

SCIENTIFIC NAME
Creobroter gemmatus

GEOGRAPHIC RANGE
Asia

Adults in this species have pale green bodies with contrasting, black-edged yellow-to-white oval spots on their forewings, which help break up the body outline and enhance their resemblance to flowers—an effective form of background matching and mimicry.

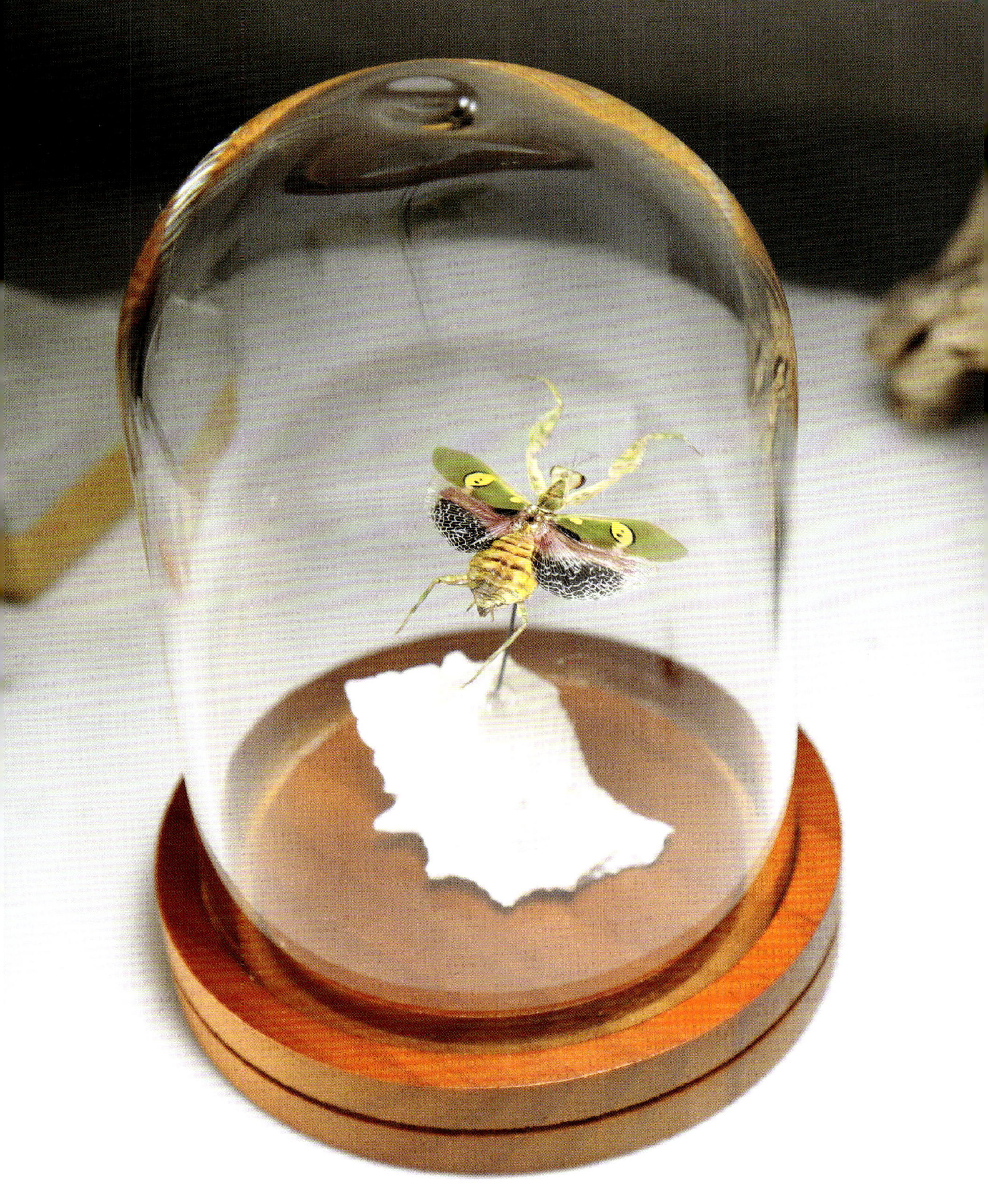

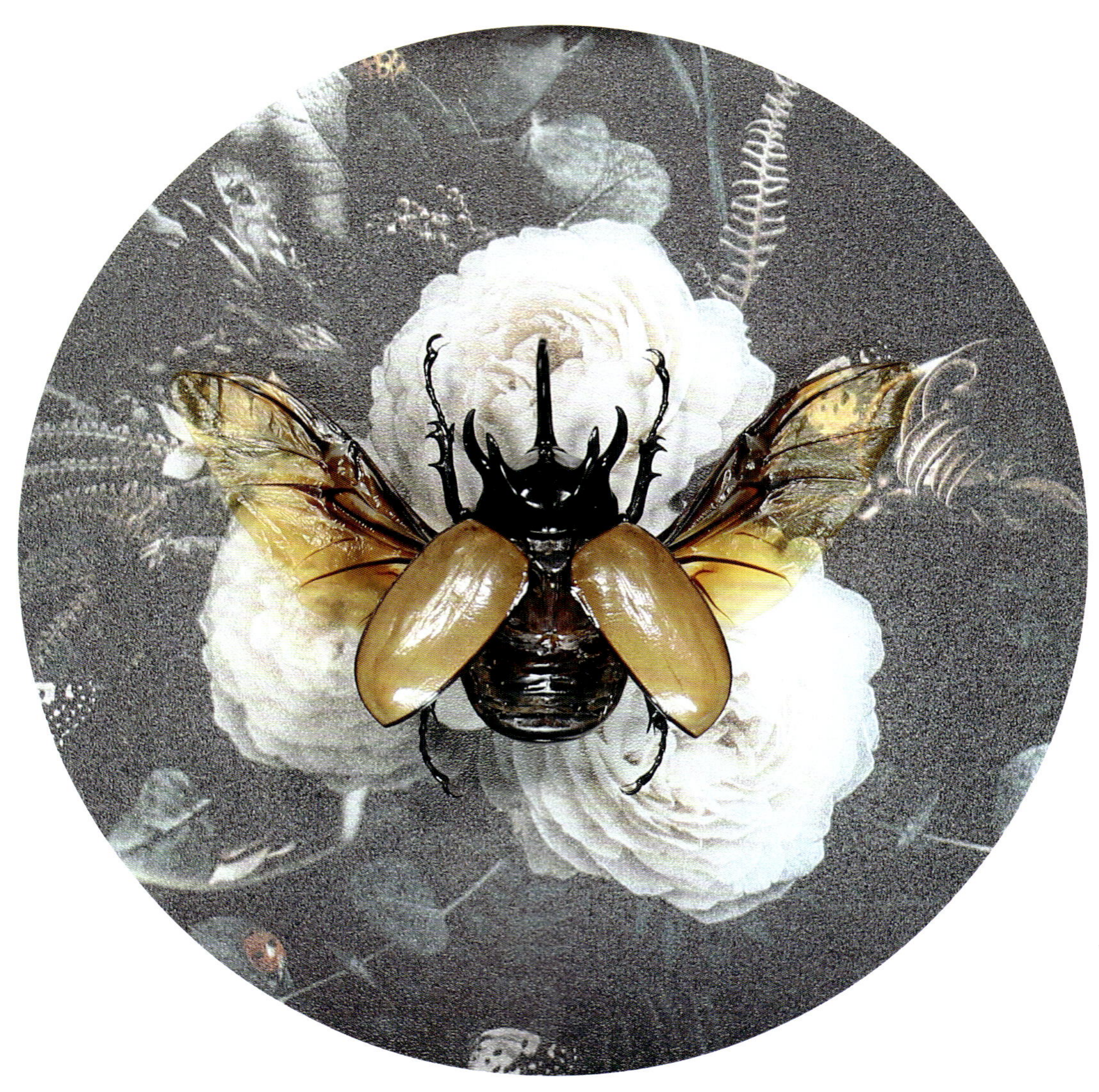

FIVE-HORNED RHINOCEROS BEETLE

SCIENTIFIC NAME
Eupatorus gracilicornis

GEOGRAPHIC RANGE
South Asia, southern China, and Southeast Asia

The body of this beetle is covered by a thick exoskeleton that provides significant protection against its predators, which include birds, mammals, and even some invertebrates. A robust pair of hardened forewings protects a second set of delicate flight wings beneath. This specialized structure enables the beetle to fly, albeit awkwardly, as its large size and weight limit aerial agility.

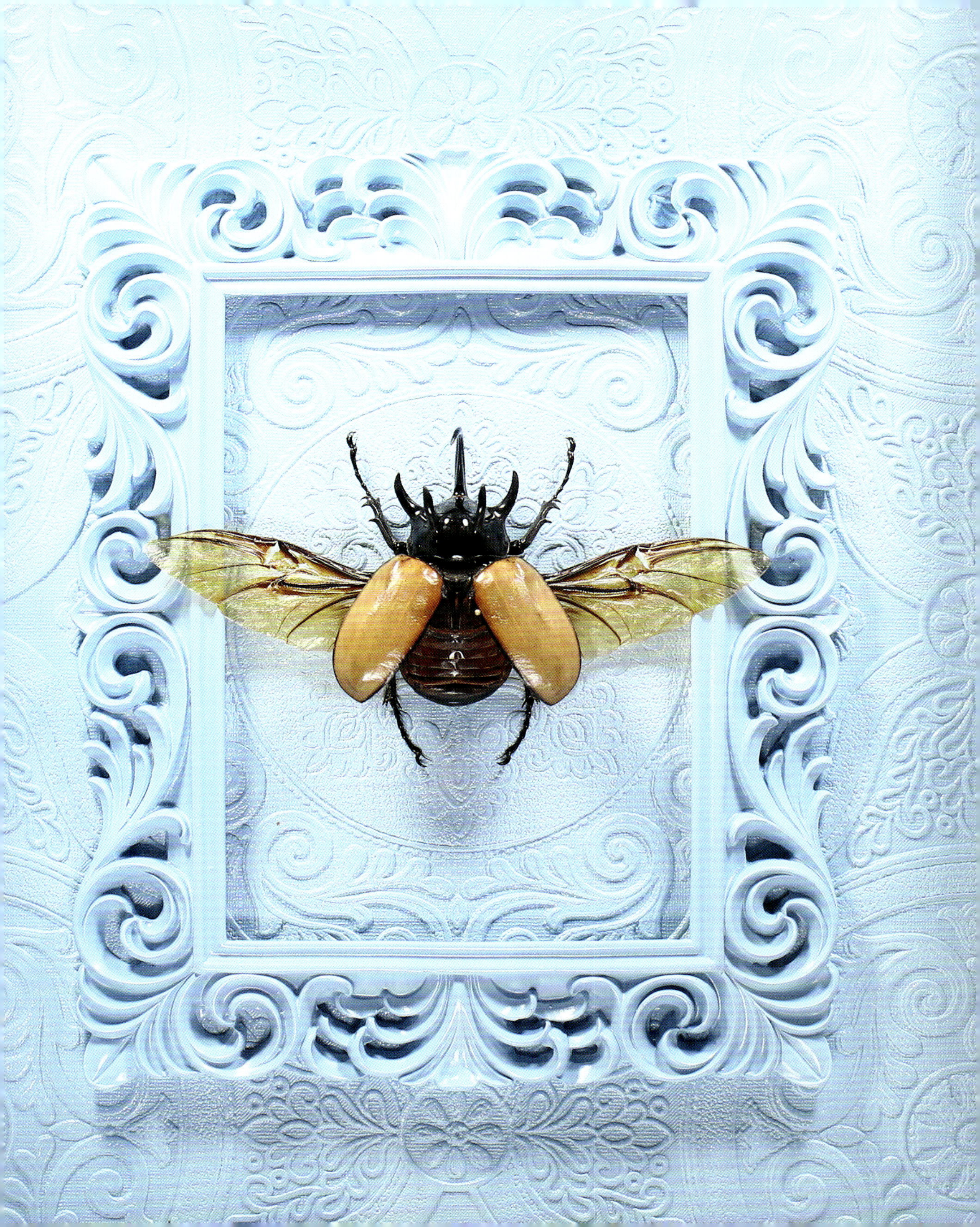

GIANT DEAD LEAF MANTIS

Similar to the ghost mantis, this mantis's leaflike resemblance provides camouflage from potential predators and aids in its hunting strategy, allowing it to ambush unsuspecting prey with surprising speed and agility. The name *Deroplatys desiccata* is derived from Latin and Greek roots: *Deroplatys* combines *dero*, meaning "neck," and *platys*, meaning "flat," while *desiccata* translates to "dried" or "dehydrated." Together, these terms describe the mantis's distinctive morphology.

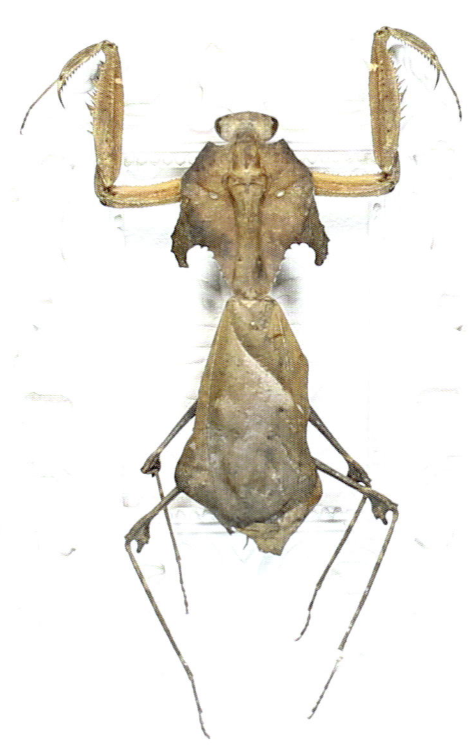

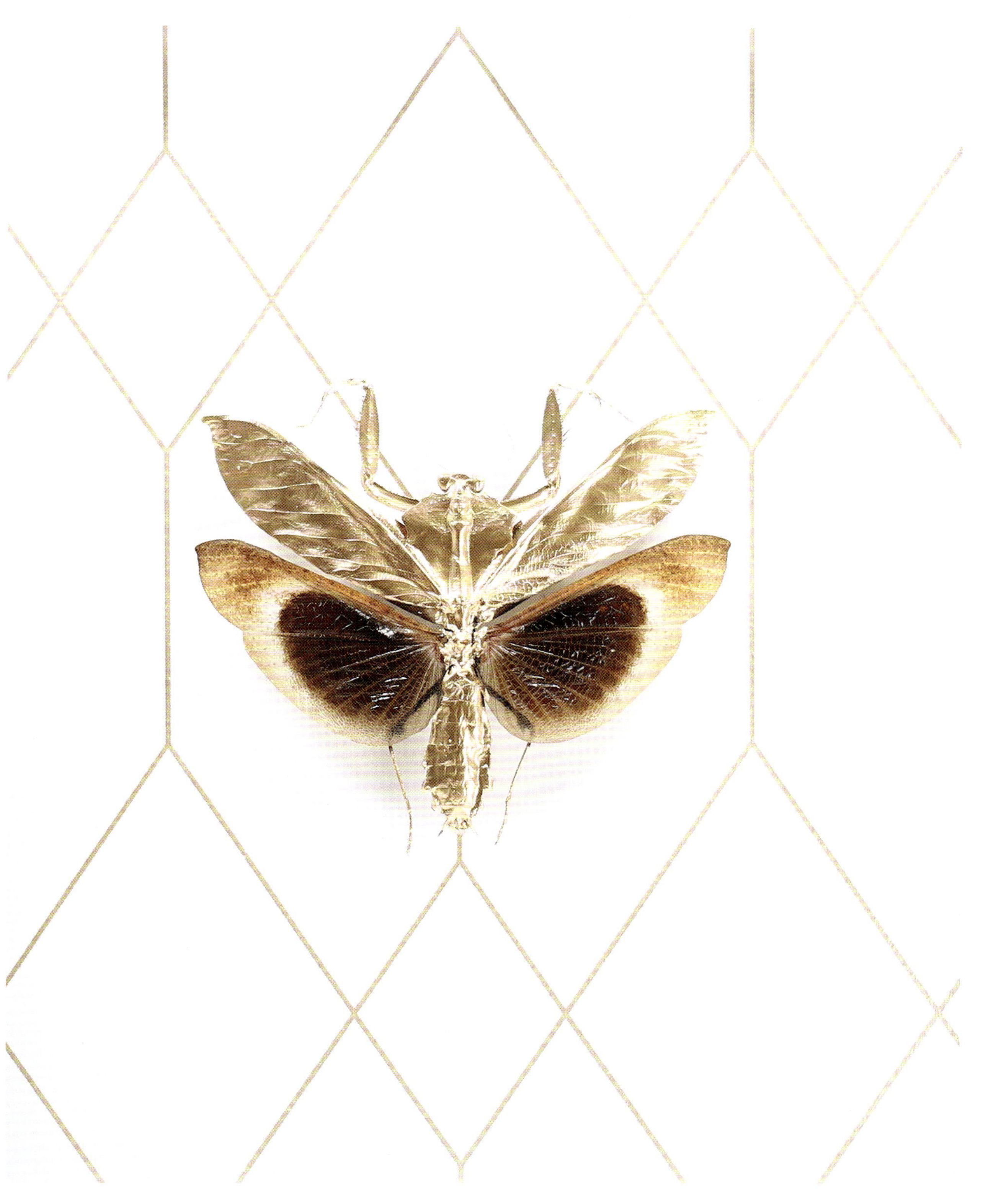

GIANT GOLDEN ORB WEAVER

Scientific Name
Nephila pilipes

Geographic Range
South Asia, Southeast Asia,
East Asia, and Oceania

This orb weaver, known for its unique appearance and intricate web-building abilities, is the second largest among the various species of orb weavers. The females are significantly larger and often more robust than the males, which are as little as one-tenth the size of females. Some of these spiders exhibit distinctive bands on their legs and bodies that have the unusual ability to reflect ultraviolet radiation. This reflection can significantly enhance their visibility to UV-sensitive prey, which may, in turn, attract a variety of insects seeking food.

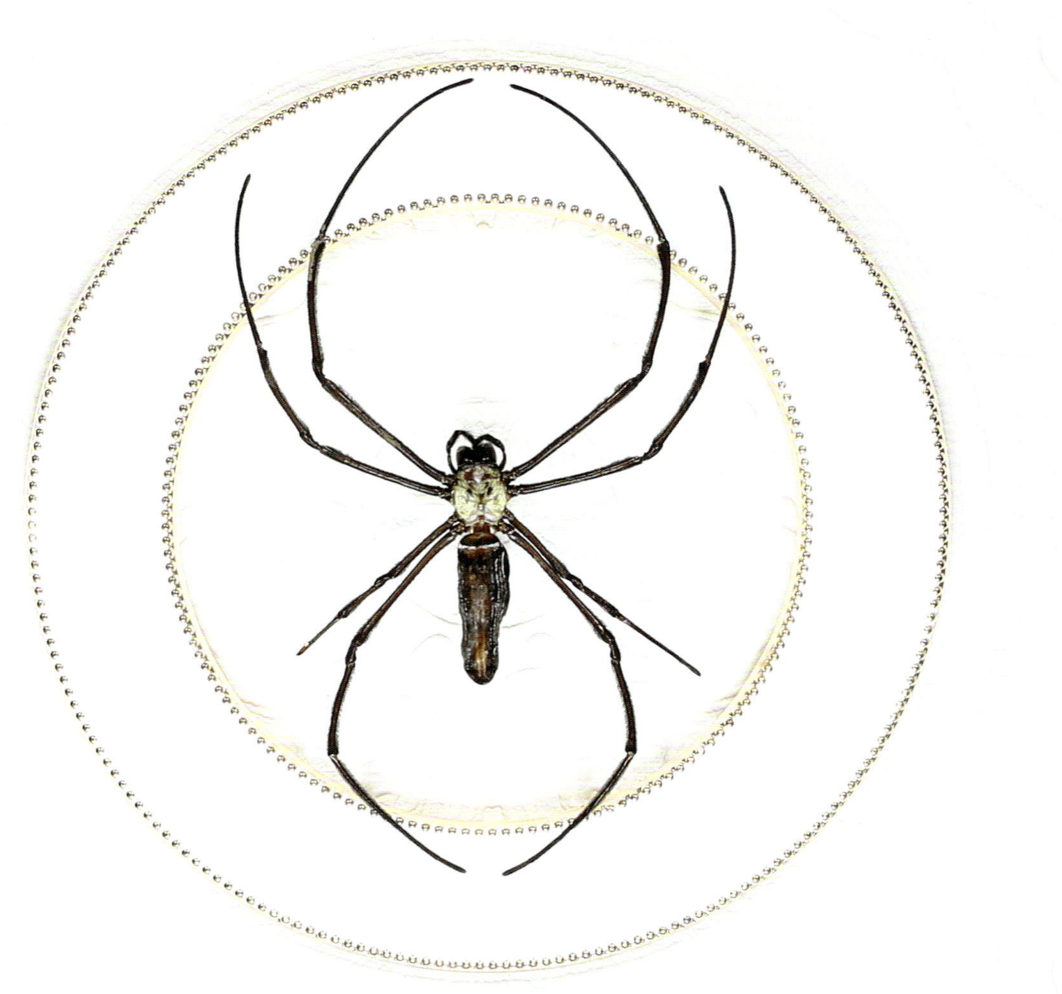

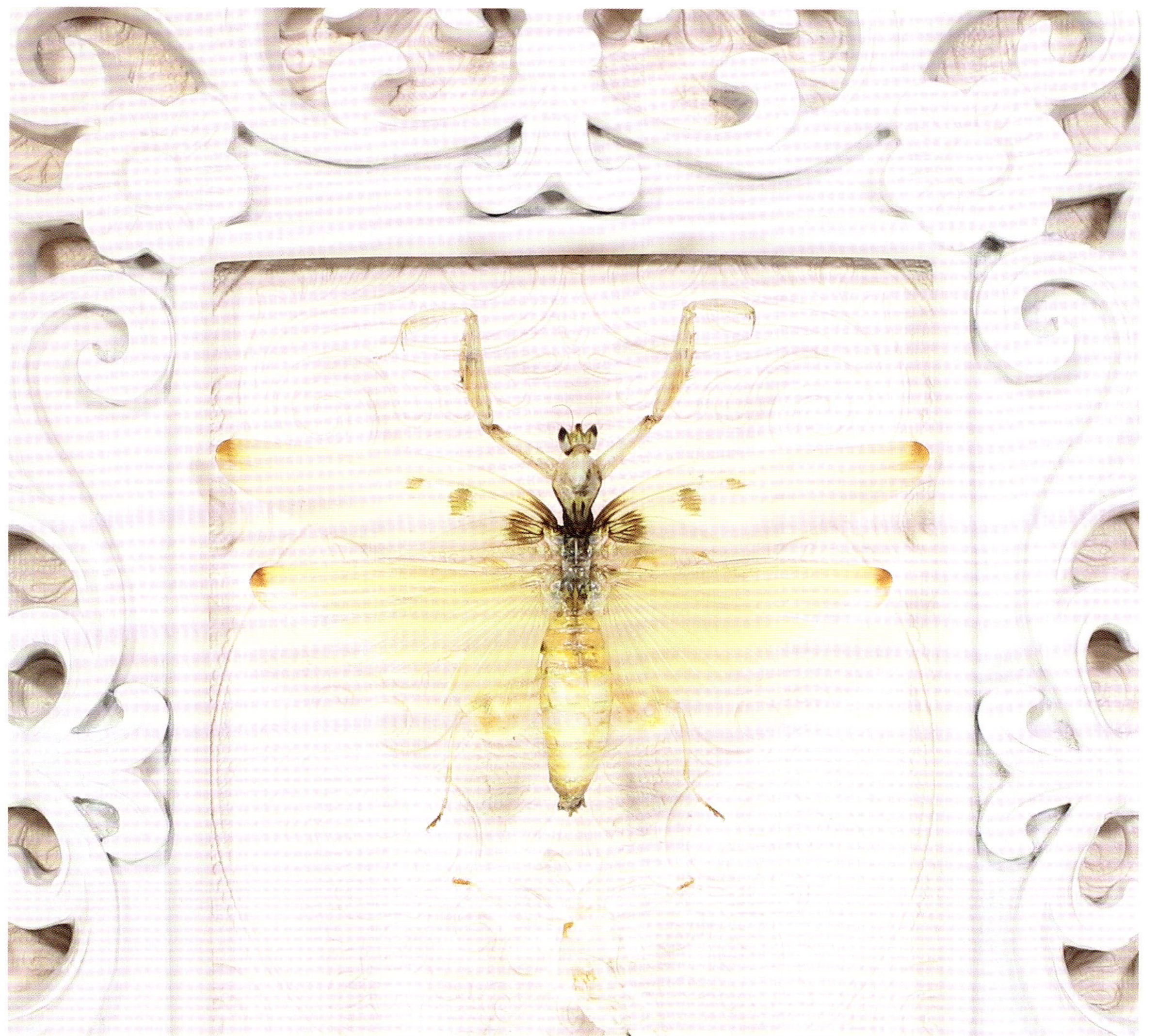

ORCHID MANTIS

SCIENTIFIC NAME
Hymenopus coronatus

GEOGRAPHIC RANGE
Southeast Asia

This stunning mantis has evolved in an extraordinary way, perfectly mimicking the appearance of an orchid flower—a remarkable example of visual deception in nature. In its juvenile stages, it showcases a vibrant pink hue that is both eye-catching and rare, making it highly prized by collectors and insect enthusiasts. Even more fascinating is its ability to lure prey in environments without flowers—it produces a scent that mimics the fragrance of the very blooms it imitates.

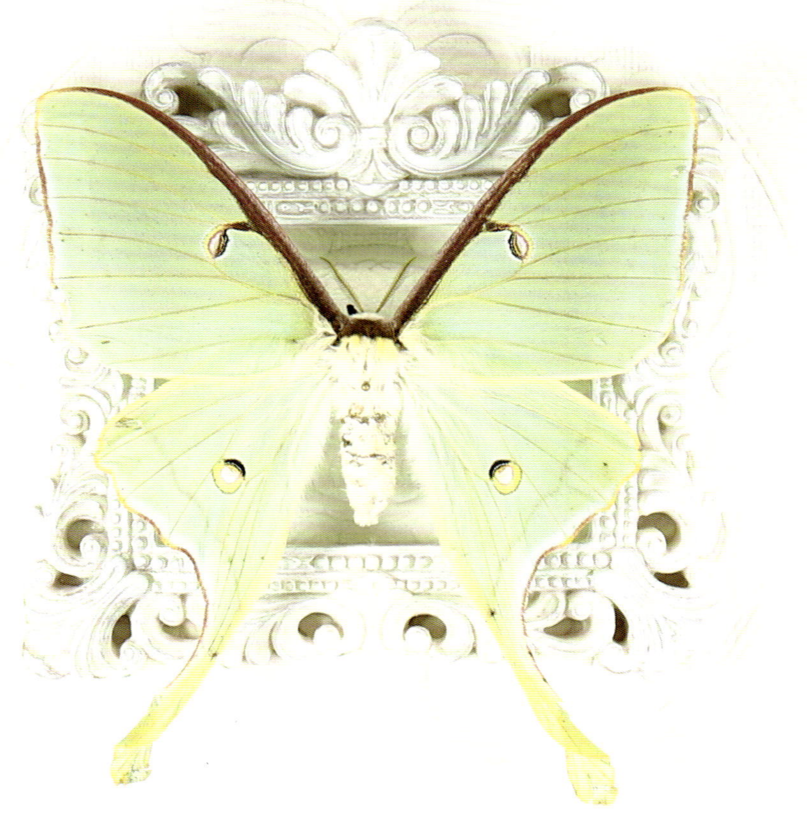

LUNA MOTH

SCIENTIFIC NAME
Actias luna

GEOGRAPHIC RANGE
North America

Luna moth larvae have two effective defense mechanisms. They can make a series of clicking sounds that warn off potential threats, and they can regurgitate their intestinal contents, which has a deterrent effect on predators. Additionally, the elongated tails of their hind wings are believed to confound the echolocation detection employed by predatory bats, making it more difficult for these bats to locate and capture their prey.

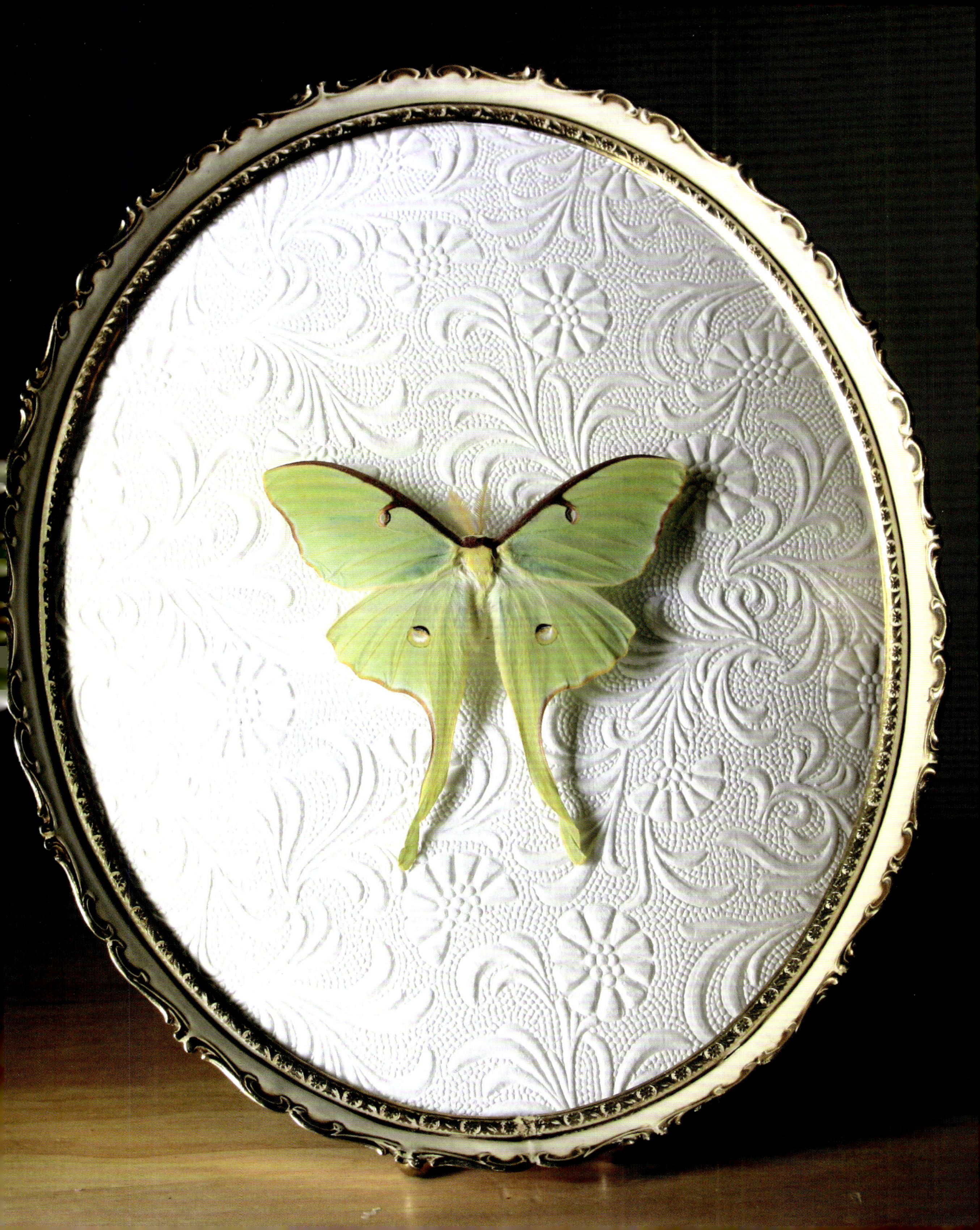

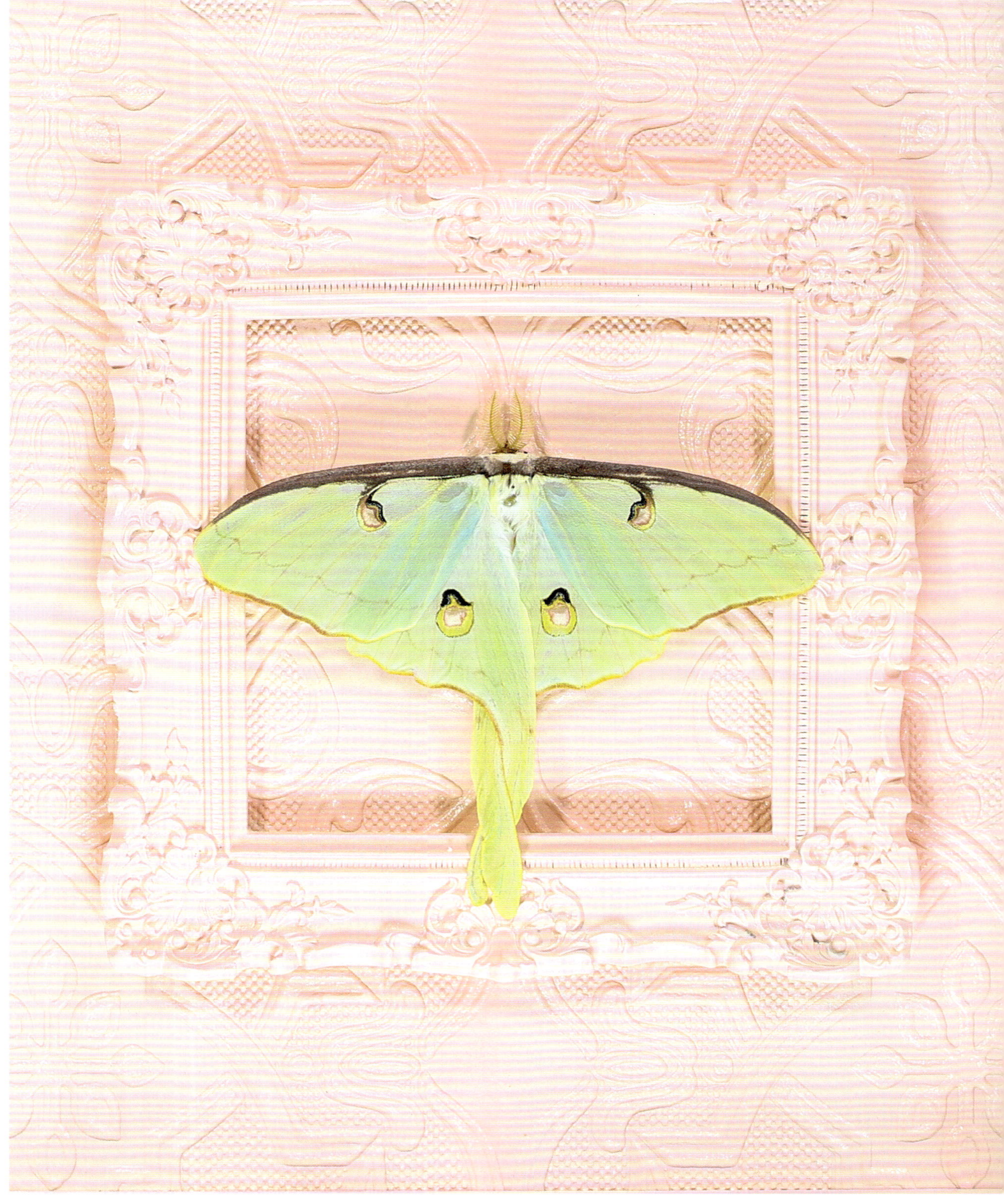

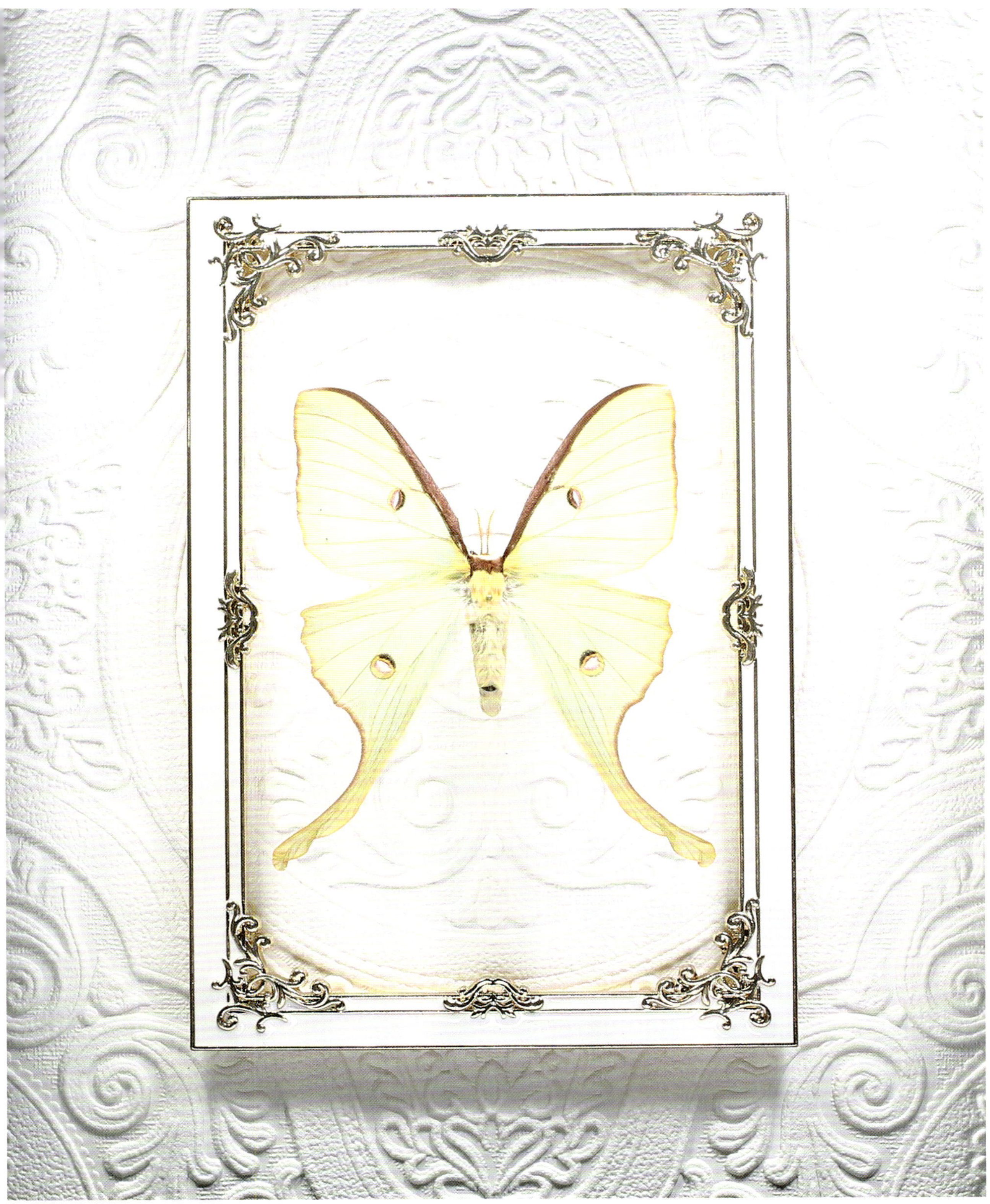

GRAY'S LEAF INSECT

The camouflage of these remarkable creatures has so thoroughly evolved that they mimic the natural movement of leaves in the wind and sway gracefully in gentle breezes, resembling real leaves in every way.

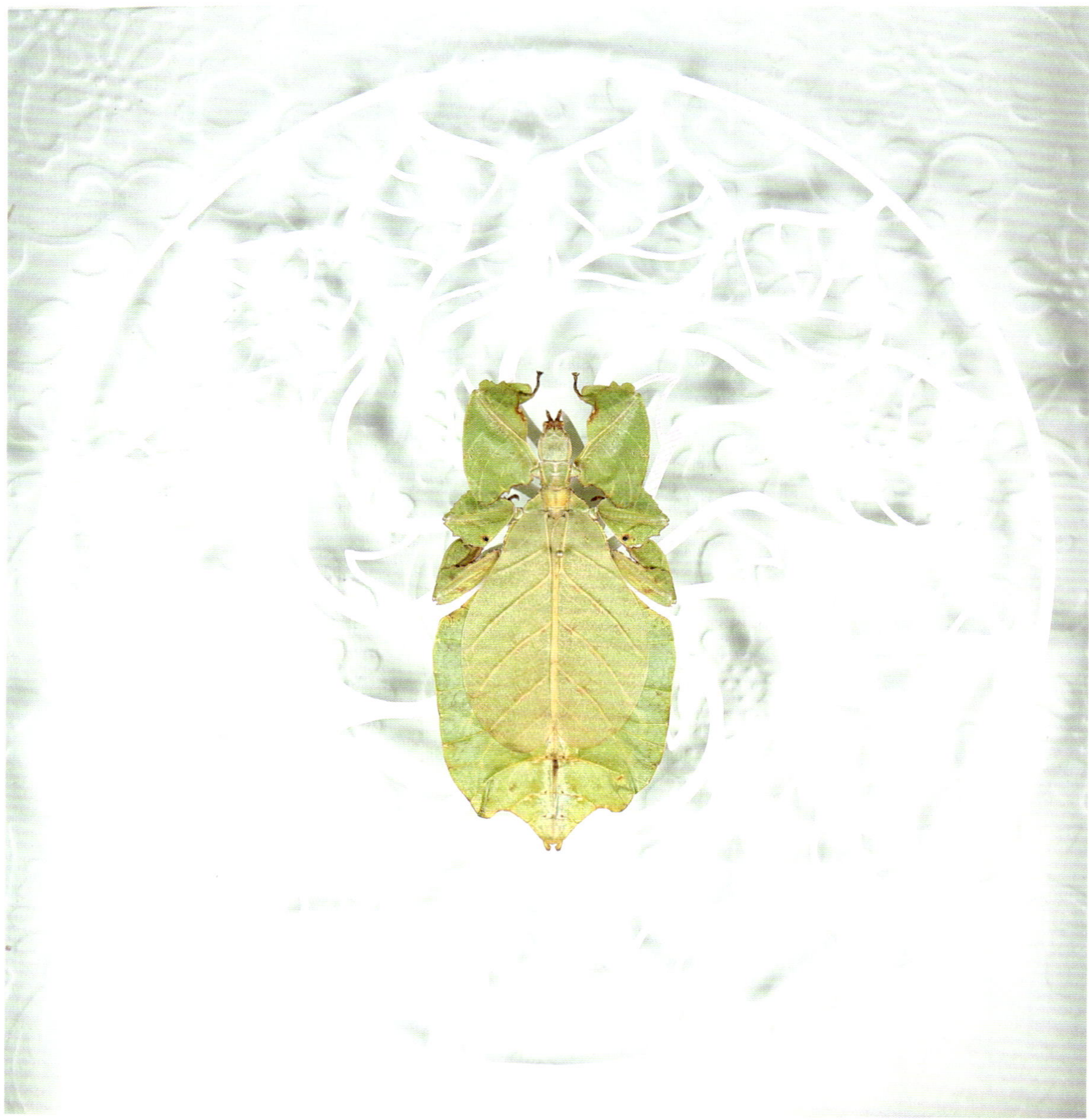

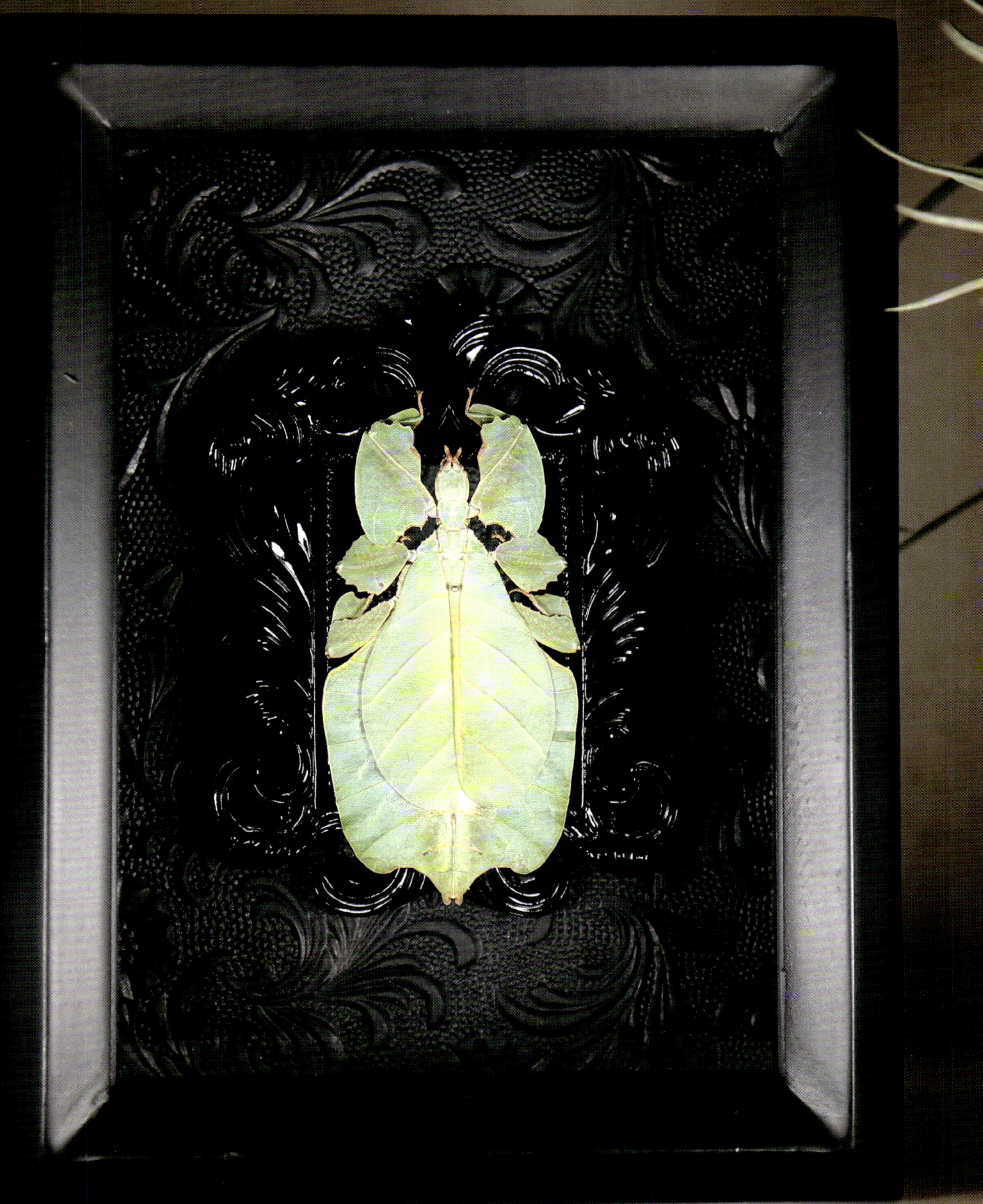

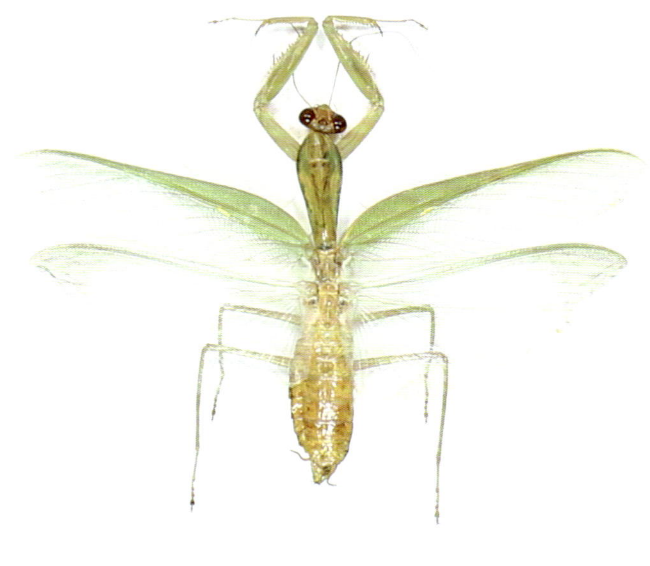

GIANT RAINFOREST MANTIS

SCIENTIFIC NAME
Hierodula majuscula

GEOGRAPHIC RANGE
Australia

The green coloring of the giant rainforest mantis helps it camouflage itself from insect predators with monochromatic (green-sensitive) vision, such as fellow mantises, but it is less camouflaged from predators like birds.

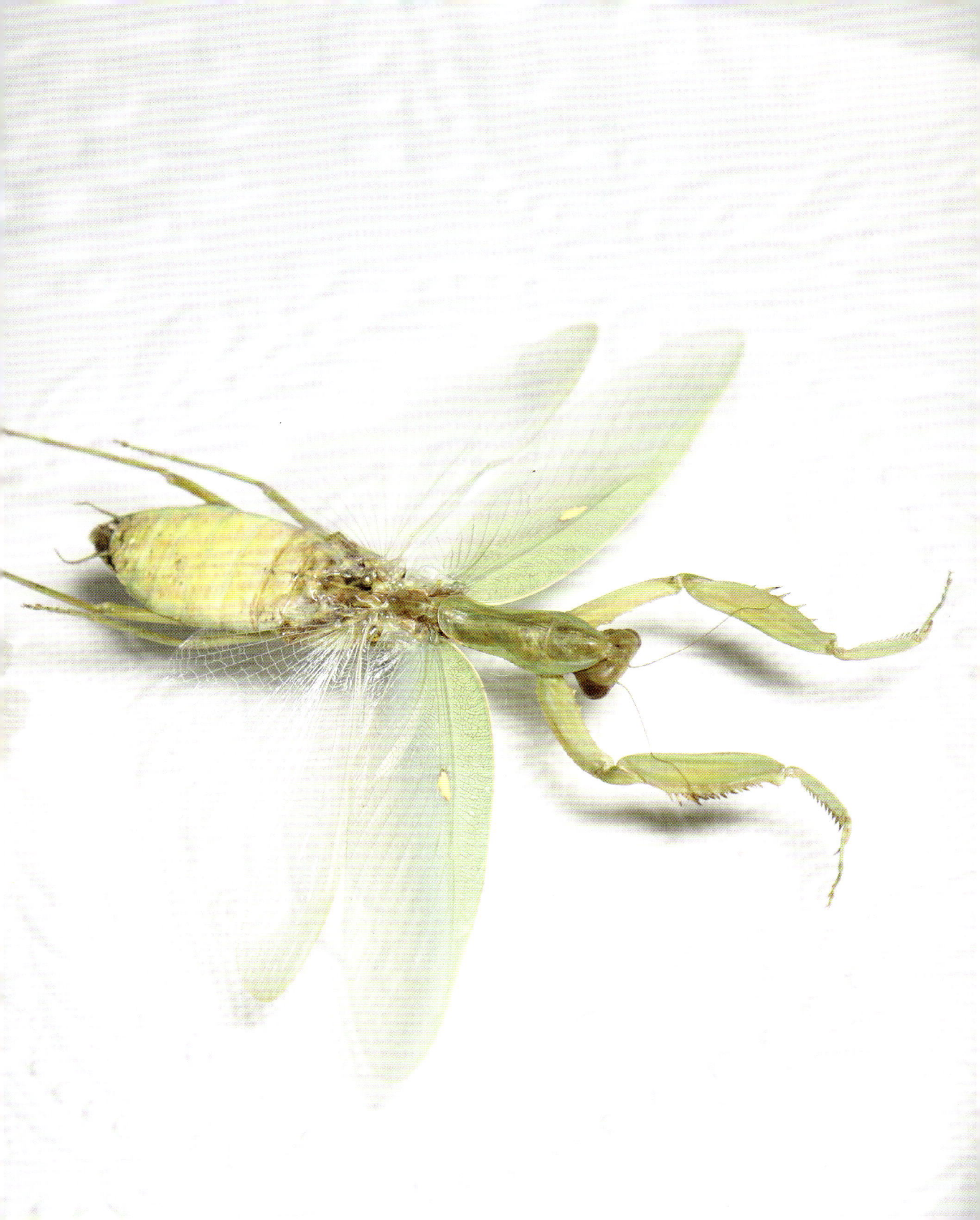

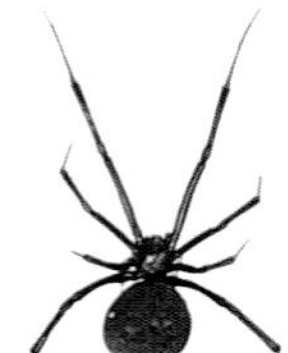

SOUTHERN BLACK WIDOW

SCIENTIFIC NAME
Latrodectus mactans

GEOGRAPHIC RANGE
North America

These infamously venomous spiders are commonly identified by the striking red hourglass marking on the underside of their abdomens. Black widow venom consists of a potent neurotoxin that can cause severe symptoms in humans, such as intense muscle cramps, nausea, vomiting, and difficulty breathing. Southern black widows are not typically aggressive and bite only in self-defense when they feel threatened. Their venom, while potent, rarely leads to death in healthy adults.

GIANT FOREST SCORPION

An individual of this species is on record as the largest scorpion in the world, with an impressive length of 11.5 inches (29.2 centimeters). Despite the giant forest scorpion's considerable dimensions, its venom is not very potent and poses little risk to humans.

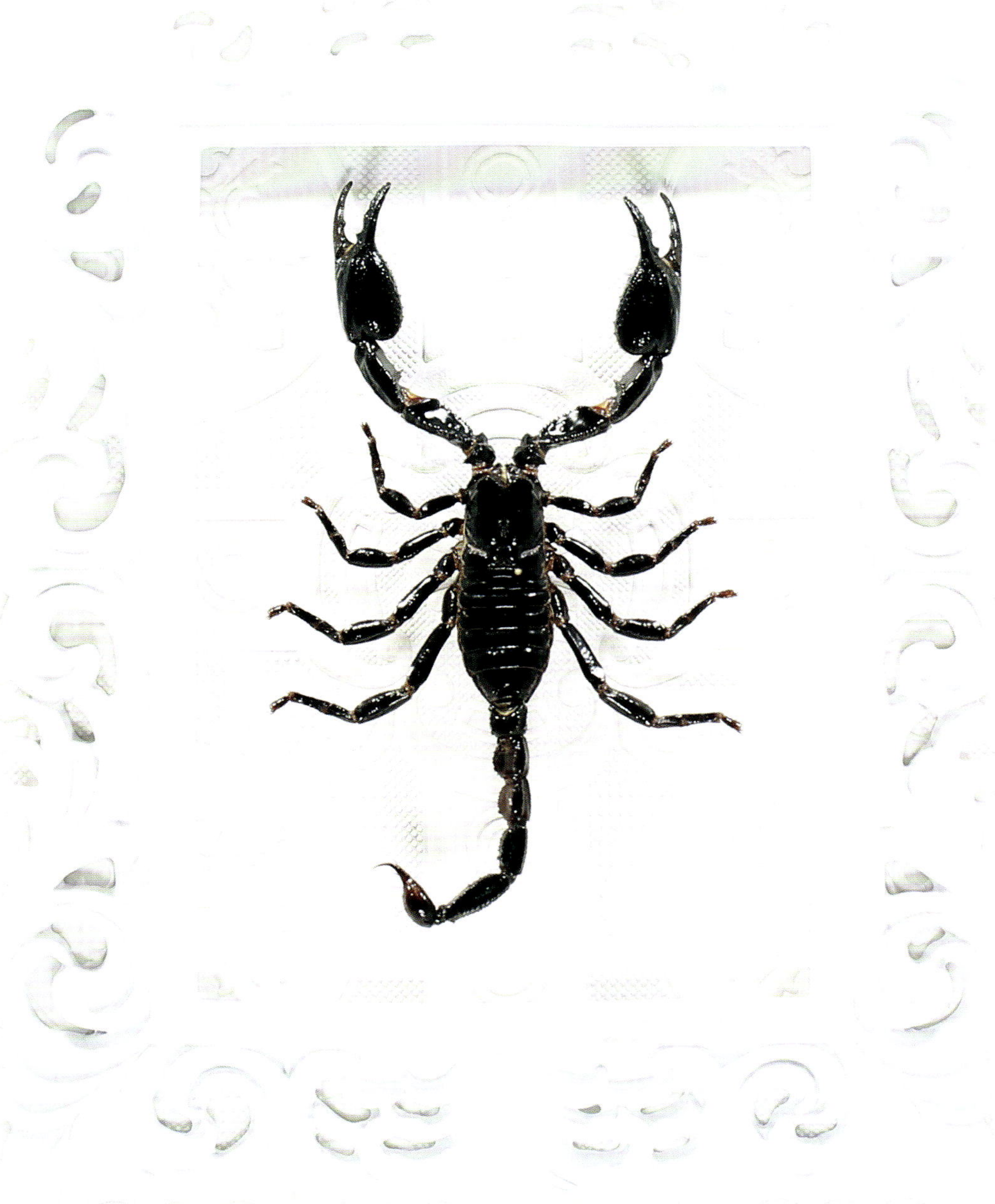

BURU GIANT STICK INSECT

198

SCIENTIFIC NAME
Anchiale buruense

GEOGRAPHIC RANGE
Buru

This species is endemic to only the island of Buru in Indonesia, where its long, slender body and narrow hind wings suggest a foliage-dwelling existence with little to no flight.

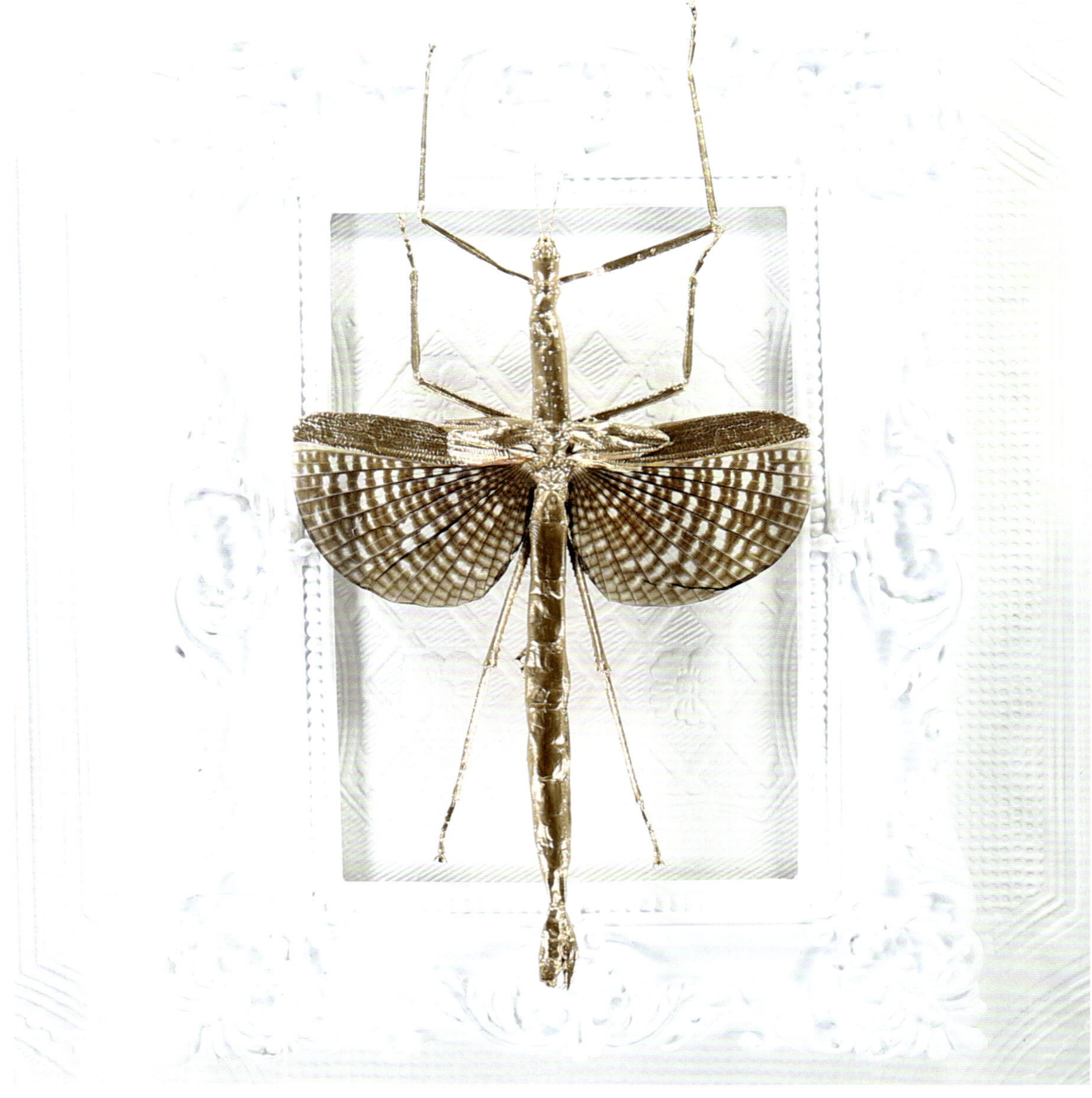

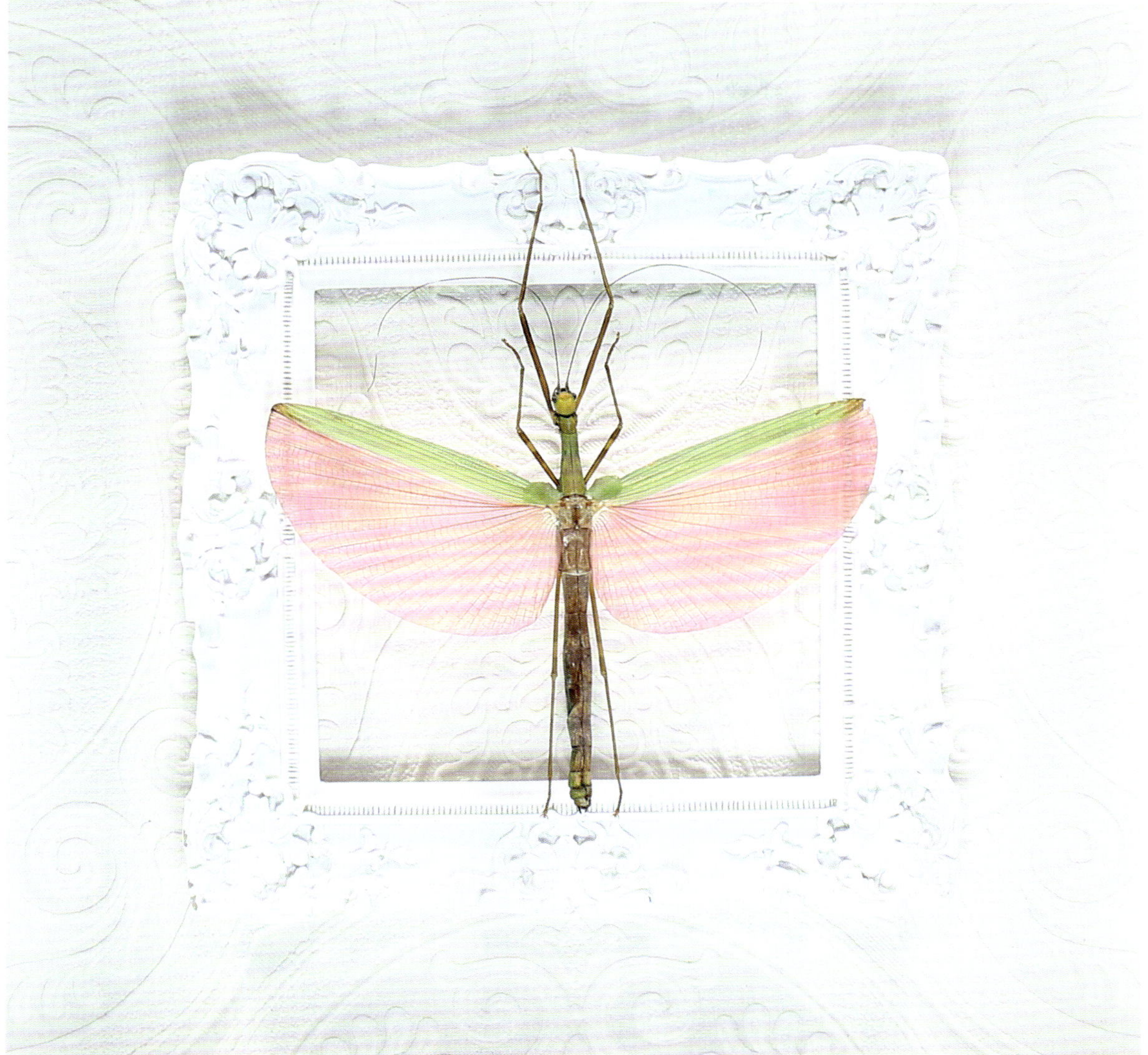

PINK-WINGED STICK INSECT

SCIENTIFIC NAME
Sipyloidea chlorotica

GEOGRAPHIC RANGE
Southern Africa,
Madagascar, Mauritius,
and Asia

The females of this species are parthenogenetic, which means they possess the ability to reproduce without mating. This reproductive strategy results in offspring that are genetically very similar or perhaps even identical to the original female—a fascinating and efficient means of population maintenance.

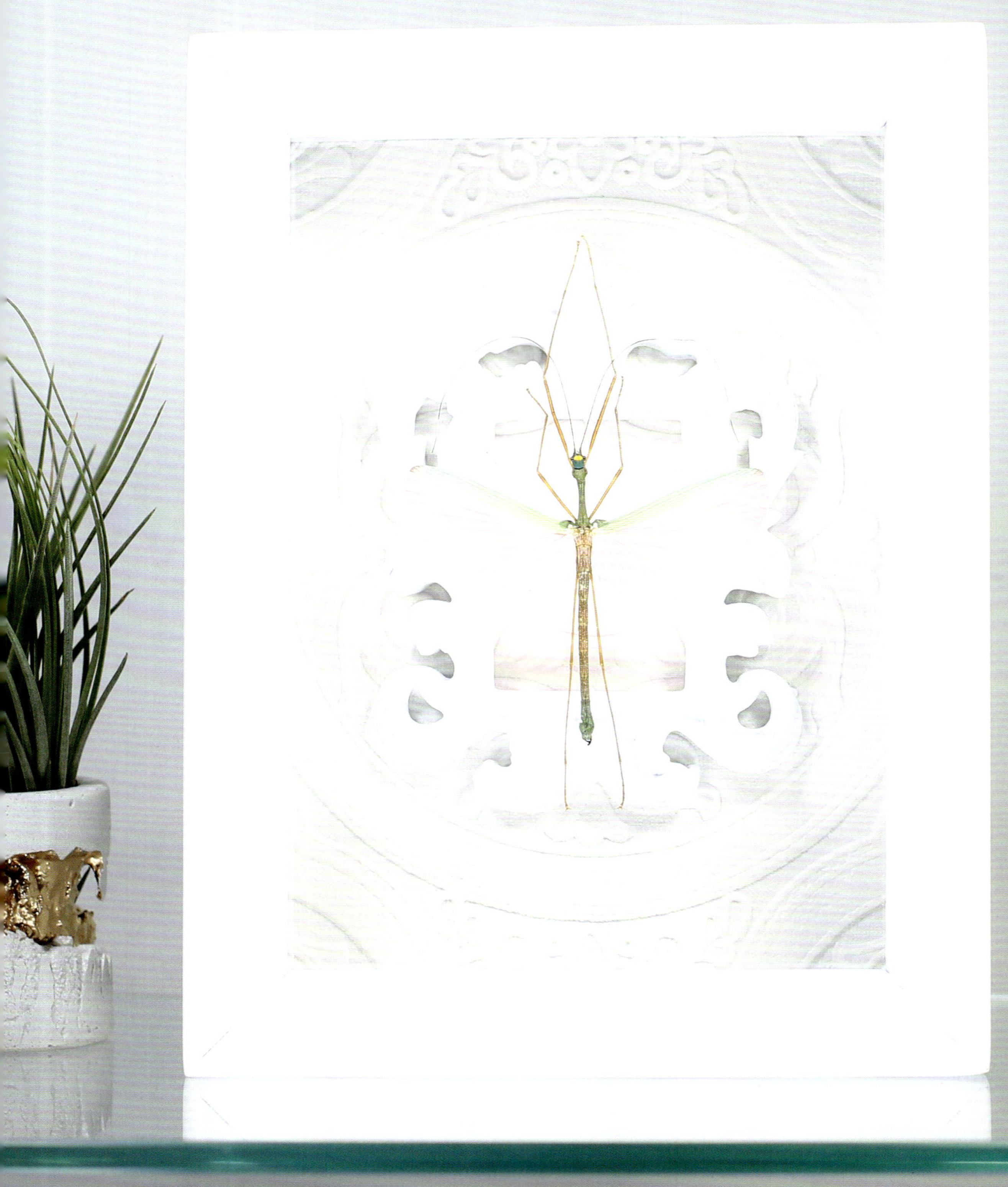

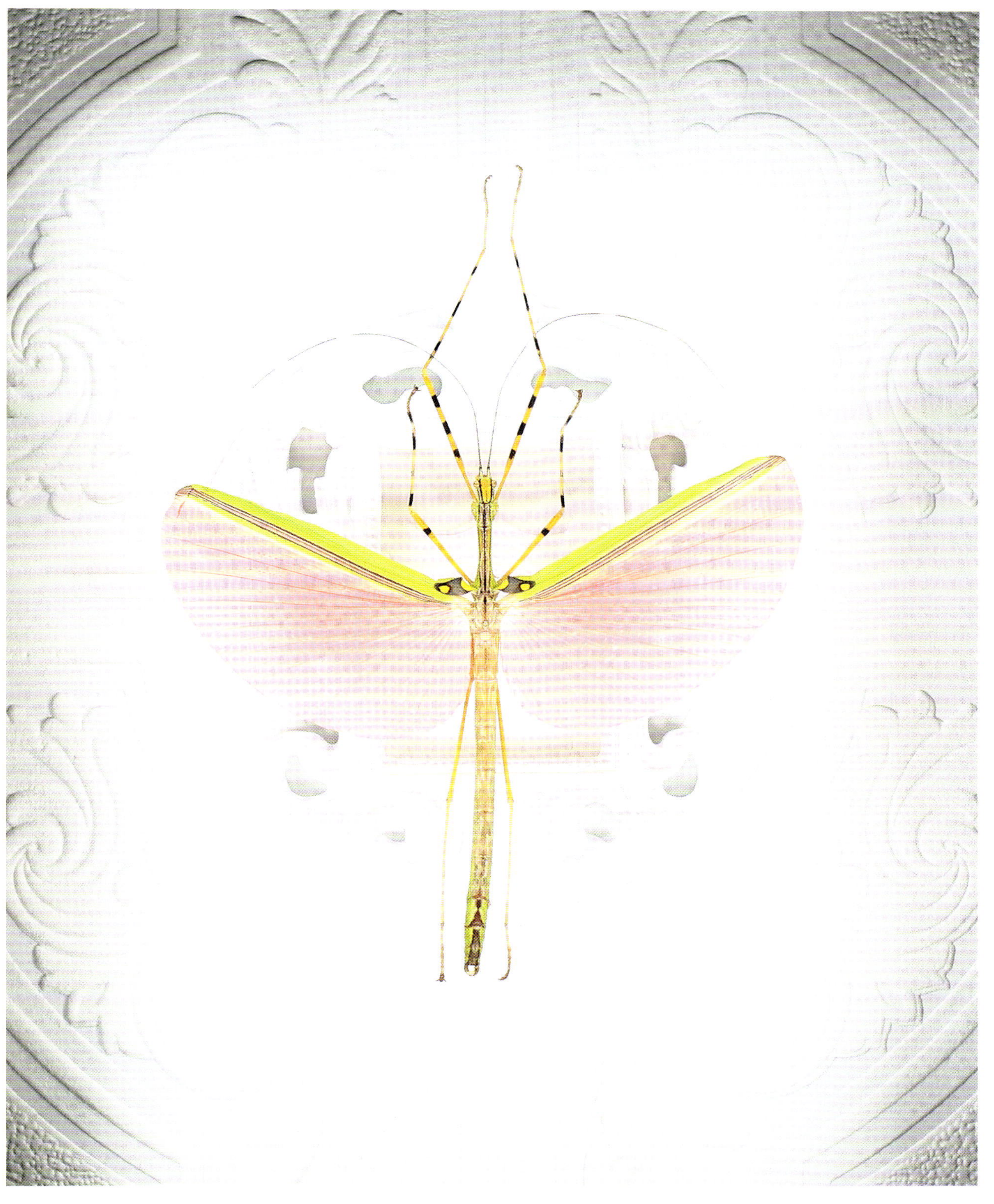

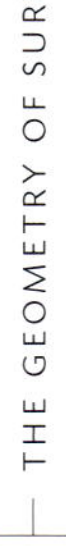

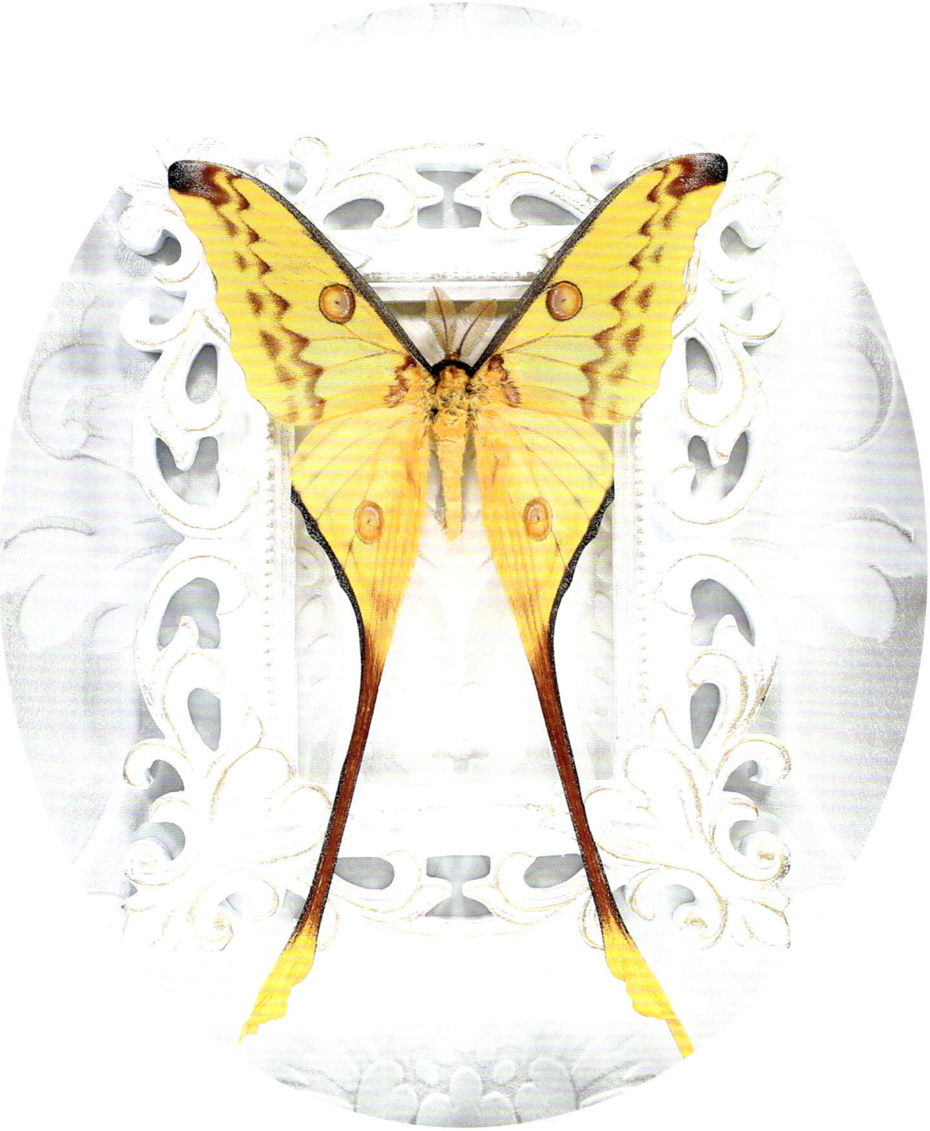

COMET MOTH

SCIENTIFIC NAME
Argema mittrei

GEOGRAPHIC RANGE
Madagascar

The comet moth uses its long, delicate wing tails to deflect incoming bat attacks and defend itself in its natural habitat. Like many insects, these moths cannot regulate their body temperature the way some other animals can, so they rely on an interesting adaptation: They vibrate their bodies and wings to generate warmth. This behavior is particularly important during cool nights, when maintaining an adequate body temperature is key to survival.

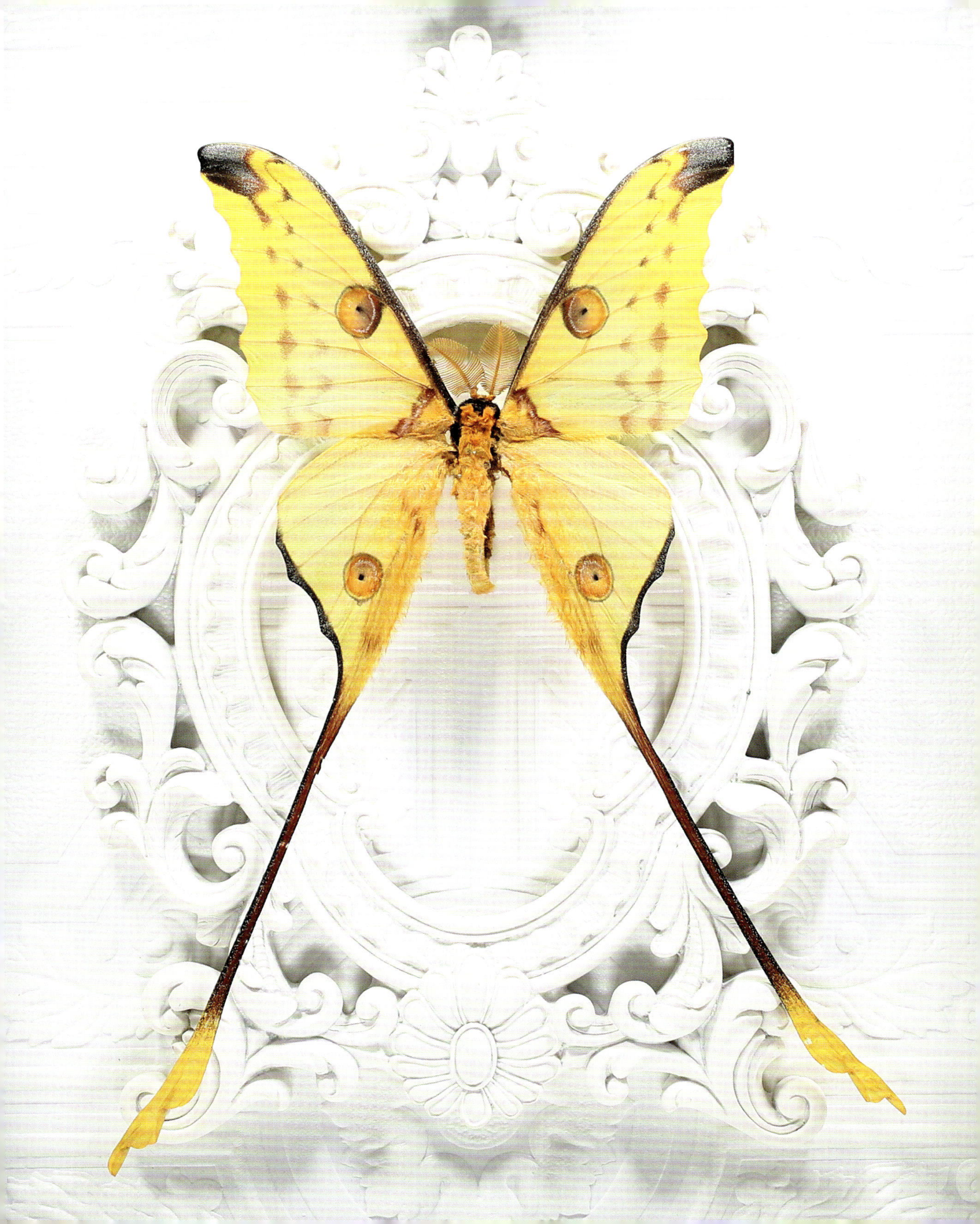

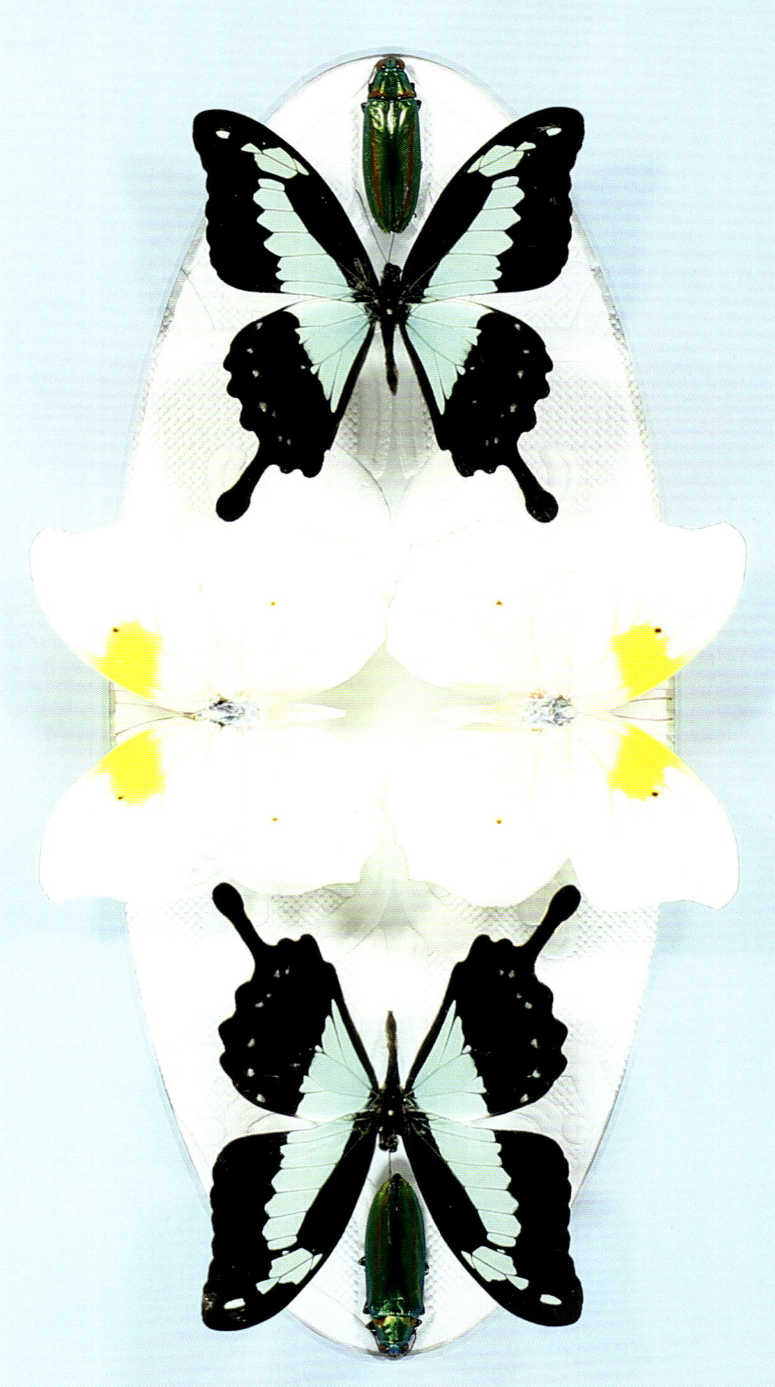

FURTHER READING

For more information, we suggest consulting the sources we used in the research of this book.

Animal Corner
animalcorner.org

Animal Diversity Web
animaldiversity.org

Annual Reviews
annualreviews.org

Australian Museum
australian.museum

Breeding Butterflies
breedingbutterflies.com

Bristol Zoo
bristolzoo.org.uk

BugGuide.net
bugguide.net

Butterflies and Moths of
North America
butterfliesandmoths.org

Butterflies of America
butterfliesofamerica.com

Butterflies of Ecuador
butterfliesofecuador.com

ButterflyCircle
butterflycircle.com

Butterfly Conservation
India
ifoundbutterflies.org

Butterfly House
butterflyhouse.com.au

California Academy of
Sciences
calacademy.org

Centers for Disease Control
and Prevention (CDC)
cdc.gov

Encyclopedia of Life
eol.org

Florida Museum of
Natural History
floridamuseum.ufl.edu

Galapagos Conservation
Trust
galapagosconservation.org.uk

iNaturalist
inaturalist.org

Indiana Nature
indiananature.net

Insects Fandom
insects.fandom.com

Insects Limited
insectslimited.com

National Geographic
nationalgeographic.com

National Parks Service
nps.gov

National Wildlife
Federation
nwf.org

Natural History Museum London
nhm.ac.uk

NatureSpot
naturespot.org.uk

One Earth
oneearth.org

Oregon State University
oregonstate.edu

Rainforest Alliance
rainforest-alliance.org

Rainforest Trust
rainforesttrust.org

Reiman Gardens
reimangardens.com

ResearchGate
researchgate.net

San Diego Zoo
zoo.sandiegozoo.org

ScienceDirect
sciencedirect.com

Smithsonian Institution
si.edu

Smithsonian National Museum of Natural History
naturalhistory.si.edu

Smithsonian Tropical Research Institute
stri.si.edu

Species File
speciesfile.org

Tree of Life Web Project
tolweb.org

UK Butterflies
ukbutterflies.co.uk

UK Moths
ukmoths.org.uk

United States Department of Agriculture
usda.gov

United States Fish and Wildlife Service
fws.gov

United States Forest Service
fs.usda.gov

University of California
universityofcalifornia.edu

Wallace, Alfred Russel. *Darwinism: An Exposition of the Theory of Natural Selection.* London: Macmillan, 1889.

Wallace, Alfred Russel. *Island Life: Or, The Phenomena and Causes of Insular Faunas and Floras.* London: Macmillan, 1880.

Wallace, Alfred Russel. *The Malay Archipelago.* London: Macmillan, 1869.

Washington State Department of Agriculture
agr.wa.gov

Wikipedia
wikipedia.org

Wildlife Trusts UK
wildlifetrusts.org

World Health Organization (WHO)
who.int

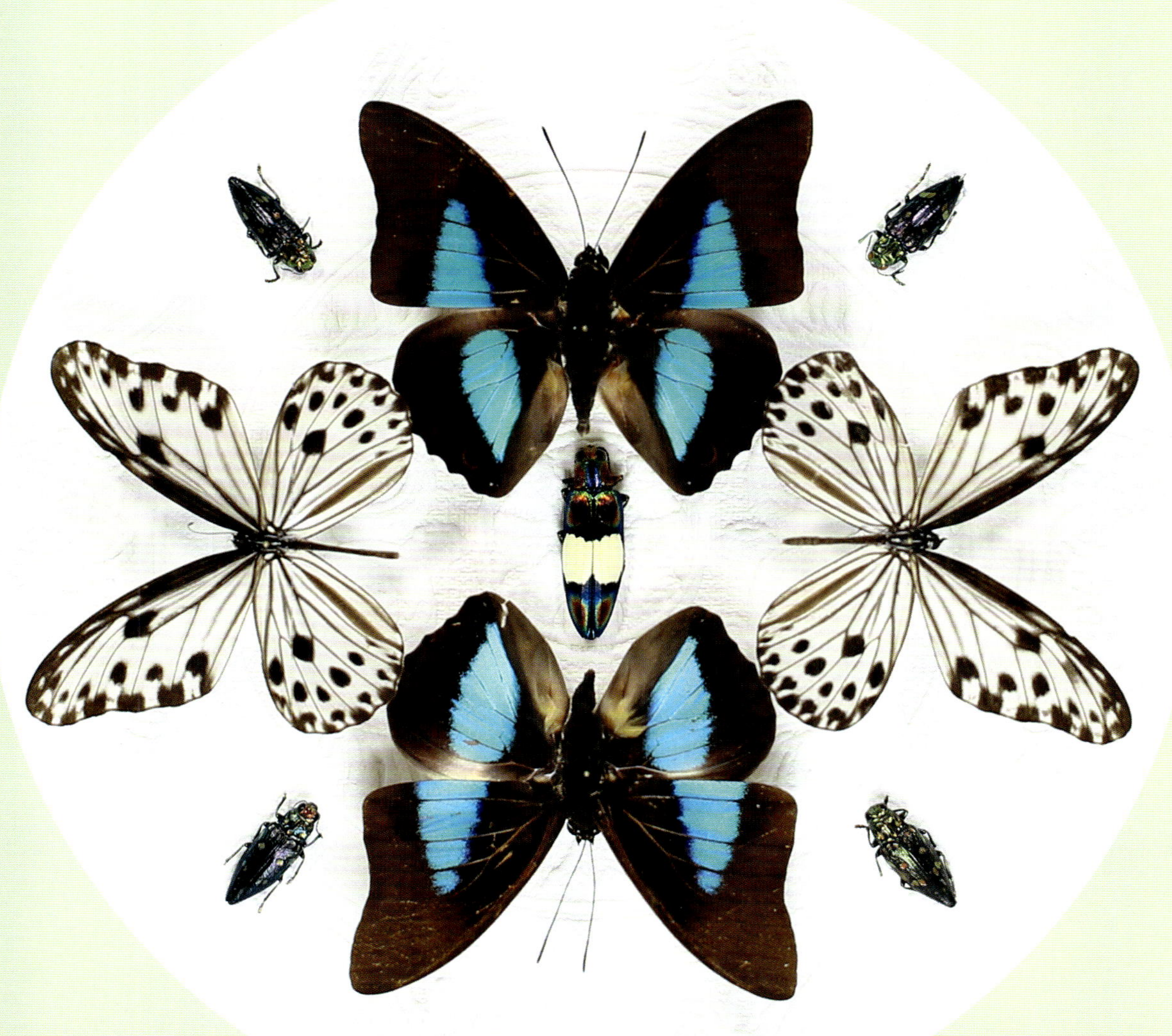

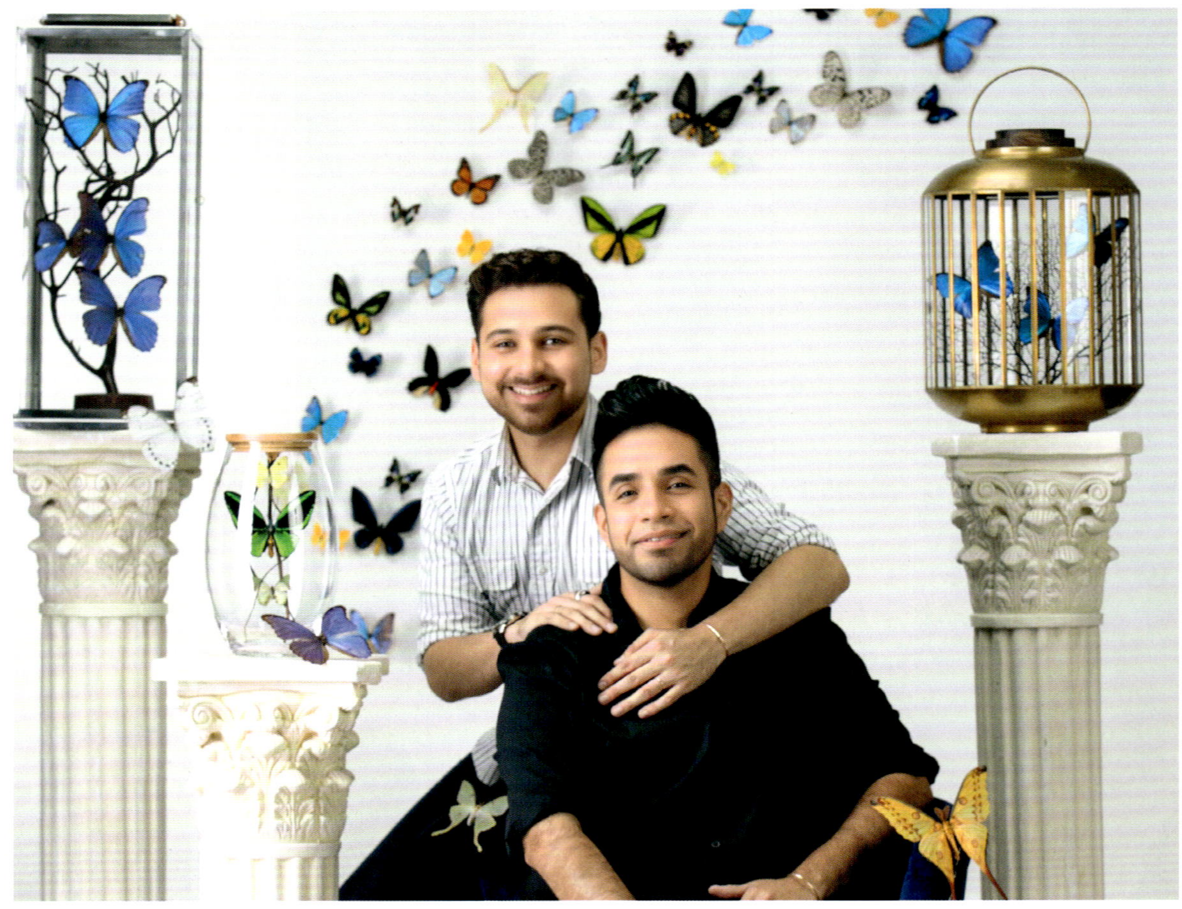

BUG IN THE BOX, a luxury decor brand based in Houston, Texas, is dedicated to the sustainable and ethical preservation of exquisite insects from around the world. Each piece in our collection transforms the natural elegance of entomology into fine art—capturing the iridescent shimmer of beetles, the delicate symmetry of butterflies, and the sculptural beauty of exotic specimens. Merging science, storytelling, and high design, we create one-of-a-kind works that celebrate nature's most intricate creations while supporting global conservation efforts.